GENES and PHENOTYPES

SERIES EDITORS
Kay E. Davies, *University of Oxford*
Shirley M. Tilghman, *Princeton University*

The identification and mapping of genes, analysis of their structures, and discovery of the functions they encode are now cornerstones of experimental biology, health research, and biotechnology. *Genome Anaysis* is a series of short, single-theme books that review the data, methods, and ideas emerging from the study of genetic information in humans and other species. Each volume contains invited papers that are timely, informative, and concise. These books are an information source for junior and senior investigators in all branches of biomedicine interested in this new and fruitful field of research.

SERIES VOLUMES
1 Genetic and Physical Mapping
2 Gene Expression and Its Control
3 Genes and Phenotypes
   Forthcoming
4 Strategies for Physical Mapping

GENES and PHENOTYPES

Edited by
Kay E. Davies
University of Oxford

Shirley M. Tilghman
Princeton University

Volume 3 / GENOME ANALYSIS

 Cold Spring Harbor Laboratory Press 1991

Genome Analysis Volume 3
Genes and Phenotypes

All rights reserved
Copyright 1991 by Cold Spring Harbor Laboratory Press
Printed in the United States of America
ISBN 0-87969-402-5
ISSN 1050-8430

Cover and book design by Leon Bolognesi & Associates, Inc.

Authorization to photocopy items for internal or personal use, or the internal or personal use of specific clients, is granted by Cold Spring Harbor Laboratory Press for libraries and other users registered with the Copyright Clearance Center (CCC) Transactional Reporting Service, provided that the base fee of $3.00 per article is paid directly to CCC, 27 Congress St., Salem, MA 01970. [0-87969-402-5/91 $3.00 + .00]. This consent does not extend to other kinds of copying, such as copying for general distribution, for advertising or promotional purposes, for creating new collective works, or for resale.

All Cold Spring Harbor Laboratory Press publications may be ordered directly from Cold Spring Harbor Laboratory Press, 10 Skyline Drive, Plainview, New York 11803. Phone: 1-800-843-4388. In New York (516) 349-1930. FAX: (516) 349-1946.

Contents

Preface *vii*

**Identification of Disease Genes on the
Basis of Chromosomal Localization**   **1**
Lap-Chee Tsui and Xavier Estivill

The Mouse *t* Complex Responder Locus   **37**
Linda C. Snyder and Lee M. Silver

**Cloning the Mammalian
Sex-determining gene, *TDF***   **59**
Peter N. Goodfellow, J. Ross Hawkins,
and Andrew H. Sinclair

**Genetic Analysis of Multifactorial Disease:
Lessons from Type-1 Diabetes**   **79**
Soumitra Ghosh and John A. Todd

Molecular Biology of the *W* and *Steel* Loci   **105**
Alastair D. Reith and Alan Bernstein

Molecular Genetics of Wilms' Tumor   **135**
Jerry Pelletier, David Munroe, and David Housman

Index *171*

Preface

It is now more than a decade since a restriction-fragment-length polymorphism (RFLP) was used to diagnose β thalassemia. This was swiftly followed by the realization that such RFLPs could be generally used to map monogenic disease. The term reverse genetics has been superseded by the term positional cloning as genes mutated in important genetic disorders have been isolated from their chromosomal position.

The genetic map of the human genome is so well advanced that if sufficient families segregating for a disorder are available, the mutant gene can be located to a small chromosomal region. More recently, the genetic map of the rodent genome has also improved with the use of microsatellite markers. The identification of mutant genes, however, can be much more difficult, particularly in the absence of an associated cytogenetically detectable rearrangement. The enormity of the task of identifying the correct candidate gene is exemplified by the continuing search for the Huntington's disease gene on chromosome 4 begun in 1983. This volume describes some examples of success stories in the isolation of human genes from their chromosomal position. Each one is instructive in highlighting the pitfalls (both methodological and psychological) of finely mapping around a gene locus and eventually identifying the correct sequence.

The search for the sex determination gene, the Wilms' tumor gene, and the cystic fibrosis gene has resulted in the development of novel and more rapid techniques for the analysis of the human genome. They illustrate that many approaches are often required and that an enormous amount of effort could be saved if a complete genome map was already available. The isolation of these three gene sequences has resulted in a detailed map of their respective chromosomal regions, which has been instructive in the comparison of both genetic and physical maps.

The three chapters on mapping and characterization of loci in the mouse demonstrate the beauty and power of molecular studies in the rodent genome. Clearly, the way forward will be comparative mapping between several organisms. This is particularly true for polygenic traits such as insulin-dependent diabetes mellitus. Such studies have implications for future analysis of other common polygenic diseases such as hypertension.

We are grateful to all the authors for their hard work and excellent contributions. Special thanks go to the staff of the Cold Spring Harbor Laboratory Press, particularly Nancy Ford and her colleagues Dorothy Brown, Patricia Barker, and Mary Cozza, who have continued in their enthusiasm to publish this series in a timely fashion. There is no doubt that the molecular basis of many more phenotypes in both humans and mice will have been elucidated during the preparation of this volume; we hope that these few examples will provide interesting reading for those inside and outside the field.

Kay E. Davies
Shirley M. Tilghman
October 1991

GENES and PHENOTYPES

Identification of Disease Genes on the Basis of Chromosomal Localization

Lap-Chee Tsui[1] and Xavier Estivill[2]

[1]Department of Genetics, The Hospital for Sick Children
and Department of Molecular and Medical Genetics
University of Toronto, Toronto, Ontario, Canada M5G 1X8

[2]Molecular Genetics Department, IRO, Hospital Duran i Reynals
Ctra Castelldefels Km 2.7 08907 Barcelona
and Genetics Service, Hospital Clínic i Provincial
Barcelona, Catalonia, Spain

The purpose of this chapter is to provide an overview of the sequence of events that on the basis of their respective chromosome locations led to the successful identification of the genes involved in some representative single-gene disorders. It is hoped that the reader will gain insight from these examples and adapt some of the strategies and techniques in studying other diseases and biological phenomena. The discussion begins with a general introduction of the subject followed by seven specific examples. In the last section of the chapter, we try to derive some common themes from our current knowledge about disease gene cloning based on chromosome position.

Specific examples discussed are:

❏ Chronic granulomatous disease was the first human disease gene to be identified on the basis of its chromosome localization. Prior knowledge of its biochemical defect made the gene identification relatively easy.

❏ Duchenne/Becker muscular dystrophy is the largest gene known in the human genome. Deletions are the most common mechanism causing the disease. There is no shortage of landmarks for delimiting the disease locus.

❏ Retinoblastoma represents a unique genetic system for studying mechanisms of somatic mutations and an excellent model for understanding the role of recessive oncogenes in tumorigenesis.

❏ Cystic fibrosis shows that it is possible to identify a disease gene solely on the basis of linkage analysis, without any chromosome rearrangements pointing the way.

❏ Choroideremia provides an extreme example in which only a little can be learned about the basic defect through cloning of the gene.

❏ Neurofibromatosis type 1 is caused by mutation in a large gene with at least three smaller genes embedded in one of its introns; the gene identification presents an application of human/mouse comparative gene mapping.

❏ Wilms' tumor demonstrates the difficulty of having too many candidate genes present in a small region of chromosome.

INTRODUCTION

Before the 1980s, the usual way to study a human genetic disease was to perform a systematic examination of tissues, cells, or body fluids from patients until a consistent biochemical observation was made. This approach provided the solution to many diseases, such as the thalassemias and some of the metabolic disorders that show a clear enzymatic defect. For other inherited diseases, the symptoms observed in patients are often too complex to allow direct deduction of the underlying biochemical defects, particularly since an observed abnormality may be a secondary or tertiary consequence of the primary lesion.

In the early 1980s, an alternative approach was introduced to permit cloning of the gene on the basis of its chromosome location. This new way of studying human disease is attractive because it bypasses the initial need to understand the basic defect of the disease but allows one to return and answer the question once the gene is cloned. Whether this approach is called "reverse genetics" (Orkin 1986), "forward genetics" (Botstein 1990), or "positional cloning" (Caskey 1991), the idea of the exercise is to isolate and characterize the genes involved in various genetic disorders and eventually to understand the respective basic defects so that more effective therapy may be devised.

The task of positional cloning can be roughly divided into three major steps: (1) chromosome localization of the disease locus, (2) identification of the gene, and (3) determination of the gene function. Although slightly different strategies are required for different diseases, with appropriate families and patient materials, localizing the genes involved in simple (single-gene) diseases to a chromosome region is probably the easiest of the three steps. For example, some diseases may be caused by gross chromosome rearrangements, such as translocations, deletions, and duplications, which serve as good markers for the position of the targeted gene. In these cases, obtaining a refined localization of the region of interest is straightforward because the region usually can be defined by careful cytogenetic examination of chromosomes from patients. Even for the majority of diseases where gene mutations are submicroscopic or due to single-base-pair substitutions, gene localization may be achieved through linkage analysis with known genetic markers, i.e., the region of interest may be defined by linked genetic markers.

The principle of linkage analysis has been known in the study of human genetics for about half a century, but the use of this method to localize a disease gene recently received wide appreciation after the introduction of DNA markers defined by restriction-fragment-length polymorphisms (RFLPs) (Kan and Dozy 1978; Botstein et al. 1980). With the more recent addition of sequence polymorphisms defined by variable number of tandem repeats (VNTR) (Nakamura et al. 1987) and dinucleotide tandem repeats (Weber 1990), obtaining a genetic map of the human genome entirely covered with DNA markers has become a realistic goal. In fact, an almost complete linkage map was constructed some 5 years ago (Donis-Keller et al. 1987). Detecting a linkage is therefore relatively straightforward, although at times tedious.

Once a chromosome location is defined, the next challenge is to delimit the critical region to a size small enough that molecular cloning experiments can be used to identify candidate genes. Since the techniques used in this step are somewhat universal, we discuss some specific examples in this chapter to provide a glimpse into the intricacy of the various applications.

Finally, with each disease studied, the most exciting phase is to understand the basic defect from the cloned gene and the mutations. Unfortunately, not only is each disease different in terms of the function of the gene product and basic defect, but there is also rapid progress in the respective fields, preventing us from providing an adequate in-depth review. The detailed implications for each of the cloned genes in the respective diseases are therefore not discussed here.

CHRONIC GRANULOMATOUS DISEASE

The first success of positional cloning was for the gene mutated in chronic granulomatous disease (CGD), which is a disorder with severely impaired host defense against infection. The basic defect was thought to be associated with the inability of the phagocytic cells in patients to generate superoxide and oxygen derivatives following ingestion of microorganisms. The disease is most often transmitted in an X-linked fashion, but an autosomal recessive form has also been reported. Several proteins (such as *b* cytochromes and flavoproteins) had been considered as sites of the basic defect for CGD (Segal et al. 1983), but biochemical studies had not been fruitful mainly because of the complexity of the oxidase system.

The identification of a Duchenne muscular dystrophy (DMD) patient (named BB) also affected with CGD, retinitis pigmentosa (XLRP), and McLeod syndrome, and with a cytogenetically detectable deletion, was crucial in the assignment of X-CGD to Xp21.1, a region of about 5000 kb (Francke et al. 1985; Kunkel et al. 1985; Baehner et al. 1986). Genetic linkage studies with DNA markers 754 and pERT84, both known to map within the BB deletion, confirmed the regional localization of CGD (Baehner et al. 1986). The precise chromosomal localization and the availability of a small interstitial deletion presumed to span the CGD locus prompted the search for the gene through knowledge of its position (Royer-Pokora et al. 1986).

Using the "complementary cloning" strategy, Kunkel et al. (1985) recovered several DNA fragments from the BB-deleted region as part of the parallel study designed for cloning of the DMD gene. The procedure is known as the phenol-enhanced reassociation technique (pERT). In brief, an excess (about 200 times) of sonicated DNA from patient BB was hybridized with DNA from a female with four X chromosomes and digested with *Sau*3A to increase the cloning efficiency of DNA from the BB-deleted region. The fragments that contained perfect *Sau*3A ends and therefore could be cloned from the mixture were those derived from the BB-deleted region and, in a relatively smaller proportion, those derived from other chromosome regions that reassociated to self rather than to the sonicated DNA (Kunkel et al. 1985). The chances of cloning DNA fragments from outside the BB deletion, i.e., reassociated fragments from the sonicated DNA and the hybrid fragments from the BB patient and the normal female DNA, were greatly reduced. The presence of phenol in the pERT procedure apparently increased the speed of reassociation of DNA fragments (by a factor of 20,000).

The clones derived from the BB-deleted region (pERT clones) were critical in the identification of both the X-CGD and the DMD genes (Monaco et al. 1986; Royer-Pokora et al. 1986). To identify sequences relevant to the CGD locus, however, filters containing the pERT clones were hybridized with a labeled cDNA probe prepared from HL60 leukemic cells that were treated with dimethylformamide for 1 week to induce the NADPH-oxidase system. In addition, a subtractive hybridization was performed with RNA from phagocytic cells of a patient (named NF), who also had a deletion in the Xp21 region and both CGD and DMD, to eliminate cDNA probes not derived from this region. The analysis yielded two overlapping genomic clones, both of which were found to detect a 5-kb mRNA species and therefore were presumed to contain expressed sequences (Royer-Pokora et al. 1986).

Overlapping clones, containing an open reading frame capable of encoding a polypeptide of 468 amino acids, were then isolated from cDNA libraries made from granulocytic cells (Royer-Pokora et al. 1986). The study of RNA from mononuclear cells from different tissues with the cDNA probe indicated that the transcript was mainly expressed in cells from the granulocytic lineage but was absent in cells from X-CGD patients. A further suggestion that this candidate was indeed the X-CGD gene came from the identification of a patient with a deletion involving part of the CGD locus. More conclusively, however, the primary sequence of the predicted polypeptide revealed that it corresponded to the β chain of the b_{245} cytochrome and was therefore in perfect agreement with the biochemical description of the disease (Orkin 1987).

DUCHENNE/BECKER MUSCULAR DYSTROPHY

DMD is a severe X-linked disease with an incidence of 1 in 3000 male births, whereas a clinically similar but less severe form of myopathy, Becker muscular dystrophy (BMD), affects 1 in 30,000 males (Emery 1987). The DMD locus was initially localized to the short arm of the human X chromosome (Xp21) by detection of structural chromosome abnormalities (Greenstein et al. 1977; Zatz et al. 1981; Verellen-Dumoulin et al. 1984; Francke et al. 1985). These structural abnormalities consisted of X-autosome translocations in affected females and a small Xp21.1 deletion in a patient (named BB) who had, in addition to DMD, CGD, XLRP, and McLeod syndrome. The assignment of the DMD/BMD locus to Xq21 was also derived from genetic linkage analysis (Davies et al. 1983; Bakker et al. 1985; Goodfellow et al. 1985).

Two different approaches were employed in the isolation of the DMD gene, but both were based on chromosome rearrangements detected in patients. In the first approach, the strategy was based on the assumption that the DMD gene was disrupted by a specific translocation in a female patient with a balanced X;21 translocation. Since the portion of chromosome 21 involved in the translocation was part of a ribosomal RNA (rRNA) gene cluster, it was possible to isolate the critical breakpoint junction fragment of this patient with an rRNA gene probe. Cloning of the junction fragment was, however, nontrivial because of the existence of a large number of tandemly linked rRNA sequences in the patient's genome. In fact, it was impossible to visualize the junction fragments in gel-blot hybridization analysis of this patient's total genomic DNA with rRNA gene probes. A crucial step in the successful isolation of the junction clone was the construction of somatic cell hybrids containing only the desired translocation junctions. With the reduced rRNA gene hybridization signal, the reciprocal translocation was visualized and one of the X chromosome junctions, XJ1.1 (DXS206), was cloned (Ray et al. 1985).

The second approach was based on the cloning of DNA located in the Xp21 region deleted in the BB patient described above (Francke et al. 1985). Using the pERT hybridization technique (see previous section on CGD), Monaco et al. (1985) obtained several genomic clones from the critical BB deletion interval. One of the pERT clones (pERT87 or DXS164) was found to be absent from the DNA of 5 of 57 unrelated DMD males and was therefore assumed to be located within the DMD locus. The chromosome walks from DXS164 spanned 38 kb of X chromosome DNA and yielded several new clones that were also deleted in the same patients deleted for DXS164. Further analysis indicated that these deletions could be as large as 250–600 kb (Monaco et al. 1985).

Additional studies with clones from the DXS164 locus in a large number (1346) of males affected with DMD and BMD showed that deletions of this region occurred in approximately 6.5% of these patients (Kunkel et al. 1986). Probes from the DXS205 region also proved to be extremely useful in the detection of deletions (Ray et al. 1985). These data, together with the long-range restriction maps constructed with probes derived from the Xp21 region (Burmeister and Lehrach 1986; van Ommen et al. 1986; Kenwrick et al. 1987), made it clear that the DMD/BMD deletions were often at least 150 kb in size, extending beyond both sides of the cloned region.

Several RFLPs were detected with DNA segments in the DXS164 region. Although no recombination was observed in the initial experiments with these probes in DMD families (Monaco et al. 1985), subsequent studies showed that recombination occurred on both sides of the locus in different families (Kunkel et al. 1986). It was therefore apparent that both DXS164 and DXS205 regions could be part of the DMD gene.

The coding sequences in the DMD region were first detected by cross-species hybridization with two genomic DNA segments deleted in a number of patients (Monaco et al. 1986). DNA sequencing of the corresponding human and mouse DNA fragments revealed two open reading frames flanked by potential 5' and 3' mRNA splice junctions. A transcript of 14–16 kb was also detected with these genomic DNA probes in fetal muscle RNA. Subsequent screening of a muscle cDNA library yielded several clones, all of which apparently hybridized to the same mRNA species. When these cDNA clones were used as probes to examine genomic DNA from DMD patients, different deletion patterns were observed (Monaco et al. 1986). Since many examples of deletions from either direction were found to terminate within this gene, the identity of the DMD gene was beyond doubt.

Overlapping cDNA clones were then isolated and characterized (Koenig et al. 1987, 1988; Hoffman et al. 1987a,b). A single, large open reading frame of approximately 11 kb was found, corresponding to a polypeptide of 3685 amino acid residues. Sequence analysis showed that this protein (named dystrophin) shared significant identity with a family of proteins of the cytoskeleton associated with the muscle cell membrane.

Genomic DNA hybridization analysis indicated that the dystrophin gene contains a minimum of 75 exons, encodes an mRNA of 14 kb, and is spread over a region of more than 2500 kb. Subsequent studies also revealed that DMD and BMD are variants of mutations in the dystrophin gene, since more than 60% of DMD and BMD cases were due to deletions of the dystrophin gene (Koenig et al. 1989). Among the deletions detected in BMD, however, it was found that most of them maintained the translational reading frame. It was therefore possible to explain the less severe clinical consequence of BMD deletions. On the other hand, most deletions that disrupted the reading frame resulted in DMD, a more severe form of the disease (Koenig et al. 1989).

Two in-depth reviews on the cloning of the dystrophin gene and the molecular genetics of DMD/BMD have been provided by Monaco and Kunkel (1988) and by Worton and Thompson (1988).

RETINOBLASTOMA

First noticed as a rare tumor of the retina in young children (1 in 20,000 live births throughout the world), retinoblastoma (RB) represents an excellent example of how study of a disease gene may later uncover a major story in control of gene expression and cell proliferation (for review, see Gallie et al. 1990). As it turned out, the gene mutated in RB, named *RB1* (although there is no evidence for a second locus), encodes a 110–140-kD phosphoprotein that binds DNA (Lee et al. 1987b) and interacts with a number of viral proteins that are known to have transforming functions (DeCaprio et al. 1988; Dyson et al. 1989; Egan et al. 1989; Whyte et al. 1989).

Mutation in the *RB1* gene is recessive in nature; therefore, it is also known as a recessive oncogene (Cavenee et al. 1983). The inherited form of the disease is, however, transmitted in families as a dominant trait, because individuals receiving a mutated allele from their parents have a high (>90%) chance of developing a tumor, due to the high frequency of a second hit (an elegant theory initially proposed by Knudson [1971]).

The chromosome localization of the *RB1* locus was first suggested by the karyotypic detection of constitutional deletions involving 13q14 found in a small number of patients (Yunis and Ramsay 1978). The demonstration of 50% enzyme activity of esterase D (ESD), a polymorphic serum enzyme marker known to map to that region of chromosome 13, in patients with a cytogenetically detectable deletion (Sparkes et al. 1980); tight linkage between *RB1* and ESD in family studies (Connolly et al. 1983; Sparkes et al. 1983); and loss of heterozygosity of other markers in that region of DNA from tumors (Cavenee et al. 1983) further confirmed the regional assignment.

Isolation of the *RB1* gene was greatly facilitated by knowledge of well-defined cytogenetic abnormalities (Turleau and De Grouchy 1987). By screening flow-sorted chromosome-13-specific libraries, additional DNA segments were isolated (Cavenee et al. 1984; Lalande et al. 1984). One of these fragments was known as H3-8 (Lalande et al. 1984). Fortuitously, the genomic DNA fragment immediately adjacent to H3-8 was found to show sequence conservation in other animal species and to detect a transcript of 4.7-kb mRNA, albeit in low abundance, in most cell types including an adenovirus-transformed retinal cell line (Friend et al. 1986). An almost full-length cDNA was then isolated (Friend et al. 1986) that was an obvious candidate for the *RB1* gene. Although cDNA clones for ESD were also isolated at about the same time (Lee and Lee 1986; Squire et al. 1986), they were too far from the region to be useful for isolation of the *RB1* gene (Lee et al. 1987a).

Two pieces of evidence provided the initial support that the isolated 4.7-kb cDNA came from the suspected *RB1* gene: (1) Genomic DNA rearrangement occurred in this gene in 30% of the tumors examined (Dryja et al. 1986) and (2) no mRNA was detectable in the tumors (Friend et al. 1986). Although subsequent studies showed that almost all RB tumors (including an original one used in the initial report) produced *RB1* transcripts (Goddard et al. 1988), the detection of additional mutations in the gene provided the support that *RB1* was the correct gene involved in RB (Friend et al. 1986, 1987; Fung et al. 1987; Lee et al. 1987b; Dunn et al. 1988; Cavanee et al. 1989; Yandell et al. 1989). Finally, the identity of the *RB1* gene was confirmed by the ability of a normal cDNA to suppress tumorigenicity of RB and osteosarcoma cell lines deficient in *RB1*-encoded protein (Huang et al. 1988).

The *RB1* gene spans approximately 200 kb of DNA and contains 27 exons (Yandell et al. 1989). More than 100 different mutations have been detected in various parts of the gene, and all mutations appear to be unique (D. Yandell, pers. comm.). Mutations in the *RB1* gene have also been detected in osteosarcoma and mesenchymal tumors (Friend et al. 1987), breast tumors (Lee et al. 1988), and small-cell lung carcinoma (Harbour et al. 1988; Yokota et al. 1988), although the role of *RB1* mutations is probably in the progression of these malignancies rather than in their initiation.

Although the cloning of the *RB1* gene appeared to be straightforward, the description of loss of heterozygosity (Cavenee et al. 1983) and the two-hit hypothesis (Knudson 1971), both initially described in RB, became two important concepts in subsequent cloning studies for disease genes including the genes for neurofibromatosis (*NF1*) and Wilms' tumor (*WT1*) (see below).

CYSTIC FIBROSIS

Although all previous attempts to identify the basic defect of cystic fibrosis (CF) (for review, see Boat et al. 1989) via the biochemical route had failed, linkage studies in CF began to play a major role in the early 1980s (for review, see Tsui and Buchwald 1991). The successful identification of the gene for CF represents the first example in which a disease gene was cloned solely on the basis of genetic linkage data.

Although it is generally difficult to obtain large families with multiple affected individuals for linkage analysis of an autosomal recessive disease, the high frequency of CF, often cited as the most common severe genetic disorder in the Caucasian population (1 in 2500 live births), made it feasible to collect sufficient nuclear families for the study. Furthermore, although most of the families had only two affected children, the large sample size provided the necessary statistical confidence.

The first linkage to CF was detected with an enzyme marker PON, defined by the varied paraoxonase levels in the sera of CF family members (Eiberg et al. 1985). The chromosome localization of CF did not come from the PON linkage, however, but from the demonstration of linkage to an arbitrary RFLP marker, D7S15 (Tsui et al. 1985), and its mapping through the use of somatic cell hybrid panel to chromosome 7 (Knowlton et al. 1985). The immediate and simultaneous discovery of two other DNA markers, MET (White et al. 1985) and D7S8 (Wainwright et al. 1985), closely linked to the CF locus, turned out to be the first fortunate step in the search for the CF gene.

The relationship of MET and D7S8 to CF was not clarified until almost 2 years later. Part of the difficulty was due to the close proximity of the two markers. With an estimated genetic distance of 1–5 cM, only a few recombination events were noted between the two markers (Beaudet et al. 1986; Farrall et al. 1988; Tsui et al. 1986; White et al. 1986; Berger et al. 1987). Taking all the data together, MET and D7S8 were found to flank the CF gene. Long-range physical mapping studies with the use of pulsed-field gel electrophoresis subsequently established that these two flanking markers were only 1.6 million base pairs apart (Drumm et al. 1988; Poustka et al. 1988; Rommens et al. 1989b).

A variety of methods were employed to obtain DNA markers closer to the CF gene before the MET-D7S8 region was physically defined. The flow-sorted chromosome-7-specific library was one of the most abundant and convenient sources for random DNA segments (Scambler 1986b; Rommens et al. 1988; Melmer et al. 1990; Jobs et al. 1990). The tedious part was to identify the segments useful for CF gene cloning. With a panel of somatic cell hybrid lines, Rommens et al. (1988) mapped more than 250 DNA segments to different regions of chromosome 7 and identified 58 segments in the critical region of about 30 million base pairs known to contain the CF gene. Two of the randomly isolated DNA segments were found to be in the MET and D7S8 interval by means of this saturation mapping technique.

Another noteworthy approach is through the use of somatic cell genetics. Taking advantage of the transforming activity of a rearranged MET proto-oncogene, it was possible to capture a small region (6 Mb) of human chromosome 7, which happened to include the CF gene, in a mouse cell line using the chromosome-mediated gene transfer technique (Scambler et al. 1986a). On the basis of the observation that undermethylated CpG-rich sequences are frequently associated with 5'ends of genes, it was then possible to isolate DNA segments enriched in these sequences from this cell line (Estivill et al. 1987b). This method led to the isolation of the IRP gene (*INT1L1*) (Wainwright et al. 1988), which turned out to be only 150 kb from the subsequently isolated CF gene (Rommens et al. 1989a). Although it did not directly result in the cloning of the gene in the case of CF, this approach can be extremely powerful in obtaining DNA sequences in critical regions where a selectable marker can be identified or inserted (Dorin et al. 1989).

The strong allelic association (linkage disequilibrium) detected around the CF locus (from MET to D7S8) signified a rather unusual region in the human genome. The association was first taken as evidence that a single mutation could account for the majority of the CF chromosomes. In addition, the extent of association for each marker was exploited in delimiting the region to be examined for the disease locus (Estivill et al. 1987a,b; T.K. Cox et al. 1989; Kerem et al. 1989).

The eventual cloning of the gene for CF was accomplished through a series of chromosome walking and jumping experiments, detection of cross-species hybridization, identification of an undermethylated CpG-rich region, and screening of different cDNA libraries made from tissues that are affected in CF patients (Riordan et al. 1989; Rommens et al. 1989a). Although 280 kb of DNA was covered by this search, only several transcribed sequences were noted. Besides the gene for CF, the other bona fide gene in this region is the IRP locus, which also associates with an undermethylated CpG-rich region at its 5'end (Estivill et al. 1987b).

The CF gene, which is now known as the cystic fibrosis transmembrane conductance regulator (*CFTR*), spans approximately 230 kb of DNA and contains 27 exons (Zielenski et al. 1991). The initial evidence that the *CFTR* gene was in fact the gene mutated in CF came from DNA sequence analysis of cDNA clones from CF patients (Riordan et al. 1989). The comparison revealed a specific 3-bp deletion in exon 10 of the clones from one CF patient, and through oligonucleotide hybridization analysis, this mutation was found in 68% of the CF chromosomes in that study population (Kerem et al. 1989).

To show that this 3-bp deletion is a CF mutation was difficult because the deletion only removes a single-amino-acid residue, phenylalanine, at position 508 (hence named ΔF508). It could easily be regarded as a sequence variation, and since strong allelic association occurred over a long distance around the CF locus, ΔF508 might be associated with some of the CF chromosomes. Using an extensive family, however, Kerem et al. (1989) showed that none of the 200 normal chromosomes studied, including some haplotype-matched, harbored the mutation. It was therefore argued that ΔF508 was in fact a CF mutation.

The assumption that *CFTR* is the long-sought gene for CF is now beyond any doubt, since more than 100 different mutations have already been detected in this gene in the remaining 30% of CF chromosomes (Cystic Fibrosis Genetic Analysis Consortium, unpubl.). Had the initial study chosen to sequence another CF mutant allele, the extensive family and DNA marker haplotype analyses would not have been necessary.

Finally, results of DNA transfection studies (Drumm et al. 1990; Rich et al. 1990; Anderson et al. 1991; Kartner et al. 1991; Rommens et al. 1991) showed that *CFTR*, a membrane glycoprotein of about 170,000 daltons, could confer a cAMP-regulated chloride channel in a variety of cell types, suggesting that *CFTR* is the channel itself. This assumption was further supported by two recent studies in which mutations in the presumptive transmembrane regions appeared to alter ion permeability (Anderson et al. 1991), and deletion of a portion of the R domain created a leaky chloride channel (Rich et al. 1991). These findings explained the electrophysiological observations indicating that the basic defect in CF was located in the regulation of chloride permeability in secretory epithelial cells and further confirmed the identity of *CFTR*.

CHOROIDEREMIA

Choroideremia (or tapetochoroidal dystrophy, TCD) is a common form of X-linked blindness characterized by progressive dystrophy of the choroid, retinal epithelium, and retina (Goedbloed 1942; Waardenburg 1942). It affects between 1 in 300,000 and 1 in 400,000 persons (Lewis et al. 1985).

Since choroideremia is an X-linked disease, the regional mapping of the TCD locus was not difficult. By means of a study of a series of RFLPs on the X chromosome, a close linkage was detected between TCD and the marker DXYS1 located in region Xq13-q21 of the long arm (Nussbaum et al. 1985). The finding was soon confirmed with additional families and other DNA markers (Jay et al. 1986; Schwartz et al. 1986; Lesko et al. 1987; Sankila et al. 1987). However, precise localization of

TCD was hampered by the relatively low recombination frequencies observed for this part of the chromosome (Drayna and White 1985; Lesko et al. 1987).

Fortunately, further data on regional mapping of TCD were derived from studies of patients with interstitial deletion and translocations. Two unrelated mentally retarded males who were diagnosed as having chromosome deletion syndrome were found to have deletions spanning Xq13-21.3 where TCD was mapped (Schwartz et al. 1986; Hodgson et al. 1987). A patient with a mild form of the disease and a balanced X;13 translocation was also described (Siu et al. 1990). Many additional deletion patients, especially those with classic TCD, were later detected and characterized with the use of DNA probes (Cremers et al. 1989b, 1990c; Wright et al. 1990). These chromosome abnormalities played a significant role in the subsequent cloning of a candidate gene for TCD (see below).

To obtain additional DNA markers from the TCD region, the pERT technique (Kunkel et al. 1985) was used to enrich for DNA sequences from the X chromosome region deleted in one of the patients with chromosome deletion syndrome (Nussbaum et al. 1987). The enrichment was marginal, however, since only 2 (DXS232, DXS233) of 83 clones with human DNA inserts actually mapped between the deletion breakpoints.

On the other hand, an anonymous DNA segment DXS165 originally isolated from an X-chromosome-specific library (Paulsen et al. 1986) turned out to be most useful in the subsequent cloning studies. Its utility was first noted when it was found to be deleted in patients with classic TCD (Cremers et al. 1987). With this and other markers, a series of deletion breakpoints were characterized; the extent of the deletions, which were later determined, ranged from 45 kb to several megabases (Cremers et al. 1990c). A primitive long-range restriction map was thus also established.

The DNA segment DXS165 was next used to isolate adjacent genomic DNA segments. In particular, a set of five DNA segments was isolated by chromosome jumping (Cremers et al. 1989a); these "jump" clones allowed the identification of four of eight deletion breakpoints analyzed, as well as the junction for the X;13 translocation described above. In addition, TCD was tentatively placed within a 625-kb *Sfi*I genomic DNA fragment detected by DXS165 (Cremers et al. 1989a). Three additional DNA probes were obtained by direct cloning of DNA enriched for this 625-kb *Sfi*I fragment from a human/hamster hybrid by field inversion gel electrophoresis (van de Pol et al. 1990). The latter probes helped to confirm some of the chromosome rearrangement breakpoints.

The close proximity of DXS165 to the TCD locus was demonstrated by two additional observations. First, a cosmid clone containing DXS165

(Cremers et al. 1989a) was found to detect a deletion breakpoint in one of the patients, and a 10.5-kb *Eco*RI fragment containing the junction between the two breakpoints was also isolated (Cremers et al. 1990b). Second, pulsed-field gel electrophoresis placed DXS165 within 120 kb of the X;13 translocation (Merry et al. 1990).

A 45-kb DNA segment containing DXS165 as well as most of the deletion endpoints detected in TCD was therefore cloned and examined for candidate genes (Cremers et al. 1990a). Two of the single-copy probes were found to hybridize to various vertebrate species, including chicken. They were used to screen a retinal cDNA library, and overlapping clones spanning a total of 4.5 kb were isolated.

The relevance of this candidate gene to TCD was established through analysis of DNA samples from patients carrying submicroscopic deletions (Cremers et al. 1990a). In all eight patients examined, missing portions of the genomic DNA could be identified readily with the use of the cDNA probes. Moreover, the breakpoint of the X;13 translocation described above also disrupted the open reading frame of the putative TCD gene. These observations provide strong evidence that the cloned cDNA is in fact the TCD gene.

Unlike the other disease genes discussed in this chapter, however, the deduced nucleotide and amino acid sequence of the TCD gene failed to reveal any significant identity with any known genes or proteins. Expression of the 5.4-kb corresponding mRNA also did not appear to be confined to the eye. Understanding of the TCD gene function is therefore considerably more difficult than all other disease genes so far identified through the same route.

NEUROFIBROMATOSIS TYPE 1

Neurofibromatosis type 1 (NF1), or von Recklinghausen disease, is a common severe autosomal dominant disorder with a frequency of 1 in 3000 (Crowe et al. 1956). Multiple cutaneous and subcutaneous nodules (neurofibromas), hyperpigmented patches of the skin (café au lait spots), and Lisch nodules of the iris are the most common features of the disease, appearing in more than 90% of individuals carrying the mutated gene. It is considered a familial cancer syndrome because of the significant risk (about 10%) of malignancy in affected individuals (Riccardi and Eichner 1986). NF1 has a remarkably high mutation rate (1 in 10^4; about 50% of patients have new mutations) and considerably variable expressivity (Riccardi and Lewis 1988).

Mapping of the NF1 gene to chromosome 17q was achieved by genetic linkage analysis in affected families. The linkage was first detected in 1987 with the DNA markers D17Z1, D17S71 and the nerve growth factor receptor (NGFR) gene (Barker et al. 1987; Seizinger et al. 1987). Further analysis showed that the NF1 gene was located in the region 17q11.2 (Schmidt et al. 1987; Ledbetter et al. 1989), and in a collaborative effort, the order of markers in the NF1 region was determined by multipoint mapping analysis: D17Z1-D17S116-CRYB1-D17S117-NF1-D17S57/D17S115-D17S54-D17S118 (Goldgar et al. 1989; O'Connell et al. 1989b), with the critical region narrowed to less than 3 cM in size (O'Connell et al. 1989a; Fountain et al. 1989).

Cytogenetic abnormalities played an important role in the isolation of the NF1 gene. Two independent cases of chromosome 17 translocation in NF1 have been reported: One patient was found to have a balanced chromosome 17;22 translocation (Ledbetter et al. 1989), and another had a reciprocal 1;17 translocation (Schmidt et al. 1987). In both cases, the translocation breakpoint on chromosome 17 was located at q11.2. Long-range physical mapping studies with two closely linked DNA markers (17L1 and 1F10, situated less than 600 kb apart in the NF1 region) showed that the two translocation breakpoints were only about 60 kb apart (O'Connell et al. 1989a; Fountain et al. 1989).

The next important step in cloning the NF1 locus was the identification of the human homolog for the mouse ectropic viral integration site-2 gene (*Evi-2*) on the basis of comparative mapping (Davisson et al. 1990). A great deal of interest was generated for this gene (*EVIA*) because the mouse gene was localized to the syntenic region on mouse chromosome 11 and is implicated in leukemogenesis (Buchberg et al. 1990). Moreover, the human homolog was found to map to the 60-kb interval between the two translocation breakpoints (O'Connell et al. 1990). Molecular analysis of the gene sequences in normal and NF1 patients, however, did not reveal any mutation (Cawthon et al. 1990). Two other genes, *EVIB* (Cawthon et al. 1991) and the oligodendrocyte-myelin glycoprotein gene, a cell-surface peptide with the properties of a cell adhesion molecule (Mikol et al. 1990), were identified by walking from *EVIA*, but, again, no mutation was found (Viskochil et al. 1990).

The final step of the NF1 gene isolation was achieved independently by two groups of investigators upon examination of additional transcribed sequences in the region defined by the above translocation breakpoints. For one group (Viskochil et al. 1990), the definition of the NF1 region was further delimited by three additional deletions in NF1 patients. As revealed by pulsed-field gel electrophoresis, one showed a deletion of 190 kb, another of 40 kb, and a third of 11 kb, and all of them encompassed the t(17;22) breakpoint; the 11-kb deletion did not involve any of the three genes described above. Moreover, a genomic DNA fragment (EE3.8) isolated from the 11-kb region was shown to cross-hybridize with mouse DNA, suggesting the presence of conserved sequences.

Since EE3.8 was also shown to be hybridized with an mRNA species of 11 kb (Viskochil et al. 1990), it was immediately used to screen cDNA libraries. Several overlapping human cDNA clones were obtained, and the hybridization data showed that they extended across both breakpoints. Alignment of the cDNA clones spanning a total of approximately 4.0 kb and the corresponding genomic DNA demonstrated that this gene was transcribed in the direction opposite to that of the three previously identified genes in this region. Subsequently, these genes were found to be embedded in a single large intron of the NF1 gene (Cawthon et al. 1990, 1991).

In the other study, Wallace et al. (1990b) arrived at the NF1 gene by chromosome jumping from a cosmid containing *EVIA*. Using this jump clone that was mapped near the t(17;22) breakpoint, these investigators isolated a cDNA clone with an open reading frame of 1.5 kb. They also isolated an overlapping cDNA by direct screening of the library with a yeast artificial chromosome (YAC) clone spanning the breakpoint. The latter cDNA, although shorter in size (0.8 kb), was shown to expand the NF1 locus by 250 kb in the telomeric direction. These cDNA clones were found to correspond to a transcript of 11–13 kb, and its expression could be detected by RNA-PCR in all tissues examined. Further study indicated that the transcript was disrupted by both t(1;17) and t(17;22) translocations (Wallace et al. 1990a,b).

Comparison of sequences published by the two studies (Cawthon et al. 1990; Wallace et al. 1990a,b) confirmed that both groups had obtained partial cDNA clones for the same transcript. Additional cDNA cloning experiments demonstrated that the NF1 gene encodes a protein of 2818 amino acids and stretches across approximately 300 kb, with the promoter located in a hypomethylated CpG island (D.A. Marchuk et al., in prep.).

The identity of the NF1 gene was confirmed by the identification of several apparent disease-causing mutations. First, a de novo insertion of a 400-bp DNA segment was detected within the gene of an NF1 patient (Wallace et al. 1990b). The inserted DNA was shown to be an *Alu* repeat element that apparently prevented proper splicing of the transcript (F. Collins, pers. comm.). Cawthon et al. (1990) also identified a missense mutation as well as a nonsense mutation in the gene in two NF1 patients. More intriguing, however, was the demonstration of the latter mutation in two other, unrelated, sporadic NF1 cases, suggesting a hot spot for NF1 mutations (X. Estivill et al., in prep.). These data thus constituted unequivocal proof for the identity of the NF1 gene.

The sequence of the NF1 gene suggested that the encoded protein shared a region of homology with the *ras* p21 GTPase-activating protein (GAP) and with the yeast equivalents of GAP, IRA1 and IRA2 (Ballester et al. 1990; Xu et al. 1990). Regardless of its possible function in gene regulation (a topic outside the scope of this review), the genetic analysis showed that the inactivation of the NF1 gene, rather than its activation, was responsible for the disease phenotype (Ballester et al. 1990; Martin et al. 1990). Therefore, NF1 appears to be another tumor suppressor gene (Knudson 1971, 1985), but it is unknown whether its action is dosage-dependent or whether it requires the presence of mutations in both copies of the NF1 gene. All mutations detected so far in NF1 patients would appear to inactivate or truncate the NF1 gene product, but no data have been obtained so far for the remaining allele in tumors. Moreover, the role of *EVI2A* and *EVI2B* in NF1 patients who develop malignancies is also unknown.

WILMS' TUMOR

The Wilms' tumor (WT) gene on chromosome 11 (p13) falls in a region involved in a contiguous disease syndrome known as WAGR (for review, see Ledbetter and Cavenee 1989), which includes, in addition to WT, aniridia, genitourinary dysplasia, and mental retardation. The mapping of the WT locus was therefore initially based on cytogenetic analysis of patients suffering from WAGR (Riccardi et al. 1978; Francke et al. 1979). Subsequent analysis of sporadic WTs with DNA markers (on the basis of loss of heterozygosity) narrowed the critical region within 11p13 (Fearon et al. 1984; Koufos et al. 1984; Orkin et al. 1984; Reeves et al. 1984). The most informative material in mapping the WT gene was a cell line (WIT-13) that was derived from a spontaneous WT with overlapping deletions of 11p13 on the two chromosome 11 homologs (Lewis et al. 1988).

Although the first marker mapped to the WAGR deletion region was the catalase gene (Junien et al. 1980), it was actually too far from the WT locus. A large number of additional markers were identified through various enrichment methods such as the use of chromosome-mediated gene transfer (CMGT) hybrids (Porteous et al. 1987; Bickmore et al. 1988), cloning from DNA fractionated by pulsed-field gel electrophoresis (Compton et al. 1990), and mapping of random segments from chromosome-11-specific libraries (Compton et al. 1988; Davies et al. 1988; Lewis et al. 1988; Gessler et al. 1989; Call et al. 1990).

Deletion maps based on overlapping constitutional deletions were established with a large number of DNA markers for the WAGR region (Gessler et al. 1989). At the same time, physical maps of the 11p13 region were also derived with the use of pulsed-field gel electrophoresis (Compton et al. 1988; Gessler and Bruns, 1989) and radiation-reduced somatic cell hybrids (Glaser et al. 1990b). As a result, the WT locus was limited to a region of less than 345 kb (Rose et al. 1990).

From a series of cosmids isolated from an 11p-specific cosmid library, Call et al. (1990) isolated one clone that appeared to map to the critical location and detected, with a fragment of this clone, hybridizing bands in DNA prepared from mouse and hamster and a 3-kb mRNA species from baboon kidney and spleen. The probe was then used to screen cDNA libraries from human embryonic kidney cells, embryonic kidney, adult kidney, and a pre-B cell line. As a result, overlapping cDNA clones of a total of 2.3 kb were isolated, and the predicted amino acid sequence revealed an encoded protein with a "zinc finger" domain indicative of a transcription factor. On the basis of its genetic localization, its tissue-specific expression, and its predicted function and possible involvement in kidney development, this gene (later named *WT1*) was proposed to be a candidate responsible for WT (Call et al. 1990).

In an independent study, an arbitrary DNA fragment near one of the many CpG-rich islands identified by long-range restriction mapping was used in chromosome walking and jumping experiments to isolate four other such islands spanning a total of 650 kb (Gessler et al. 1990). DNA fragments from these islands were then used to perform cross-species and RNA hybridization experiments and resulted in the isolation of a 3-kb cDNA, identical to that for *WT1*.

Evidence for a direct role of *WT1* in WT came later through the detection of small, internal deletions of the gene in the tumor samples (Haber et al. 1990; Cowell et al. 1991; Huff et al. 1991). In a third study (Huang et al. 1990), however, a transcript of 2.5 kb was identified with a CpG island probe isolated from several overlapping YAC clones from the WT region (Bonetta et al. 1990). The surprising result was that this 2.5-kb transcript (WIT-1) was found to have originated in an opposite direction within 600 bp of *WT1*, but the identity of WIT-1 was unknown.

Finally, although most of the effort in WT gene analysis has been directed toward the same region in 11p13, at least two other WT loci have been suggested by deletion mapping studies and linkage analysis (for review, see Francke 1990). Furthermore, mice found to be deleted for the murine homolog of the WT gene do not show any evidence of WT (Glaser et al. 1990a). Therefore, the WT story is to be continued.

CONCLUDING REMARKS

As shown by the above examples, the study of human disease by "positional cloning" represents a new approach; the challenge combines multiple disciplines in biological research, which includes genetics, molecular biology, and biochemistry, as well as a good working knowledge of the disease itself. The general concept, premise, and objectives are markedly different from those in conventional gene analysis, such as cDNA isolation for protein species abundant in tissues, gene cloning on the basis of sequence homology with other organisms, and genes induced in cells infected by viral agents or stimulants. Many techniques have been devised particularly for analysis of complex genomes with specific application to disease gene cloning.

One major variable in the degree of difficulty in mapping the disease locus is the nature of the underlying genetic alterations. For example, chromosome deletions and translocations in DMD and CGD, plus the X-linked inheritance, were distinct advantages that led to the rapid delimitation of the critical region for these diseases. For disease without chromosome rearrangements, one would need to rely on linkage analysis, which is relatively more tedious. In the case of CF, however, the relatively clear-cut diagnosis, the large number of families, and the fortuitous linkage disequilibrium allowed rapid mapping of the disease locus to an interval of less than 1.5 cM.

Precise chromosomal localization based on linkage analyses would be considerably more difficult for many rare or neuropsychiatric disorders for which diagnosis is arbitrary, with so-called low "penetrance," or variable "expressivity." For example, a slight uncertainty would greatly diminish the confidence of distance estimate by linkage analysis, even when the marker is part of the disease locus. As a result, family studies would not permit effective delimitation of the critical region to a size manageable by any conventional cloning procedures. Having a complete gene map of the region (from human genome projects) and systematic examination of the encompassing genes may be the only way to solve the problem. The problem of multifactorial (or multigenic) diseases is even more complex and will require strategies yet to be demonstrated.

Linkage disequilibrium has been a useful guide in the mapping of the CF locus (Kerem et al. 1989). The observed disequilibrium is due to the association of the major (70%) mutation with a rare haplotype of the closely linked DNA markers. Presumably, the same would be detected for other diseases where there is a major mutation, but it is impossible to predict at the outset. There is also no exact method to calculate the genetic or physical distance based on allelic association. If an association is detected, however, one can safely assume that the disease locus is near, and haplotype analysis can also be used to estimate the critical interval (T.K. Cox et al. 1989).

Once a disease locus is defined, identification of closely linked and flanking DNA markers is the logical next step. The introduction of flow-sorted genomic DNA libraries (Deaven et al. 1986), somatic cell hybrid panels, chromosome jumping (Collins and Weissman 1984; Poustka and Lehrach 1986), radiation-reduced hybrids (Goss and Harris 1975; D.R. Cox et al. 1989), the pERT procedure (Kunkel et al. 1985), CMGT, and chromosome microdissection techniques (Röhme et al. 1984; Lüdecke et al. 1989) all made isolation and characterization of chromosome-specific DNA markers relatively simple. In obtaining polymorphic DNA markers, the discovery of VNTR (Nakamura et al. 1987) and simple repeat sequences (Weber 1990) provides additional power to RFLP in family analyses.

Pulsed-field gel electrophoresis (Schwartz and Cantor 1984) is another important invention adapted to the mapping of the complex genomes (Smith and Cantor 1986). A practical physical map of the critical region is an essential component in positional cloning. The information provides a physical description of the size of the region and locations of undermethylated CpG islands that often mark the 5' ends of genes (Bird 1986). The need to isolate large pieces of genomic DNA also prompted the development of cloning technology with YAC (Burke et al. 1987; Burke 1991). An application has been demonstrated in the cloning of the NF1 gene (Wallace et al. 1990b) and the *WT1* region (Bonetta et al. 1990). It is anticipated that much of the chromosome walking and physical mapping will be simplified when the YAC technology becomes routine and widespread.

Identification of candidate genes in the critical region is then the next challenge. Although all of the disease genes previously identified have been based on sequence conservation among different animal species, the method is certainly neither robust nor efficient because not all coding regions are expected to be detectable by cross-species hybridization studies. Hypomethylated CpG islands have been useful in detecting promoter regions of some genes (Bird 1986), but, again, not all genes are marked by these sequences. New methods must therefore be developed to improve this critical step in positional cloning. Some strategies, such as those based on detection of splice junctions (Duyk et al. 1990; Buckler et al. 1991), seem to be one step in this direction.

Information about genes in the region of interest would provide an instant entry point to mutational analysis. One early example illustrating the power of gene mapping is the cloning of the opsin genes and definition of color blindness (Nathans et al. 1986a,b), although it is strictly speaking not "positional cloning." The mapping of the rhodopsin gene (Sparkes et al. 1986) and retinitis pigmentosa (McWilliam et al. 1989) allowed immediate identification of the gene defect (Dryja et al. 1990). The best available example is probably the localization of the fibrillin gene (Lee et al. 1991; Maslen et al. 1991) and association of a defect in this gene in Marfan syndrome (Dietz et al. 1991).

The use of the mouse/man sentenic region represents a powerful approach to identify candidate genes for human diseases. The cloning of the NF1 gene is an excellent illustration of this comparative mapping technique (Fountain et al. 1989; O'Connell et al. 1989a). The wealth of mouse genetics may also serve as a reservoir for some of the disease phenotypes that are otherwise impossible to dissect in human patients. For example, insertion mutagenesis allows one to create a similar disease in animals where the affected gene may be isolated through cloning of the insertion site (Woychik et al. 1985).

Because of uncertainty about the function of the gene being sought in most cases, the only guaranteed way to prove the identity of a candidate is through identification of mutations and demonstration of a straight correlation between mutant alleles and disease in patients. On the other hand, predictions based on the encoded amino acid sequence and tissue distribution of the gene product should provide supportive evidence and information about the function of the gene, although they might not constitute sufficient proofs. For example, prior knowledge about the biochemical defect in CGD was essential in the identification of the gene (Royer-Pokora et al. 1986). The availability of functional assays for the disease gene product, such as in *RB1* (Huang et al. 1988) and CF (Drumm et al. 1990; Rich et al. 1990), would give an unequivocal identity for a candidate gene. Without any biochemical or physiological understanding of the disease, however, it would be extremely difficult to determine the status of a candidate gene even with mutation data, such as in the case of choroideremia (Cremers et al. 1990a).

Besides direct sequence analysis of DNA cloned from patients, several methods for the detection of point mutations have been developed during the last few years to identify sequence mismatches; they include the RNase protection assay, denatured gradient gel electrophoresis (Myers and Maniatis 1986), chemical cleavage (Cotton et al. 1988), direct sequencing of DNA amplified by PCR (Wong et al. 1987), and more recently, single-strand conformation polymorphism (Orita et al. 1989). These are all convenient methods to detect sequence alterations, and choice of these methods often depends on habit and influence of neighboring or peer laboratories.

Many disease genes have been identified through positional cloning in the past several years; some of the recent genes discovered are those for sex determination (Sinclair et al. 1990), familial adenomatous polyposis (Groden et al. 1991; Joslyn et al. 1991; Kinzler et al. 1991; Nishisho et al. 1991), X-linked spinal and bulbar muscular atrophy (La Spada et al. 1991), fragile X syndrome (Verkerk et al. 1991), and the Greig cephalopolysyndactyly syndrome (Vortkamp et al. 1991). The list will likely increase substantially during the next 3–5 years. For some disorders where detailed physical maps of the critical regions have already been constructed and intensively examined by many groups, e.g., Huntington's disease (Bates et al. 1991), polycystic kidney disease (Harris et al. 1990), and myotonic dystrophy (Shaw et al. 1989; Smeets et al. 1990), we should expect to see their conclusion even sooner (Pritchard et al. 1991).

In addition, study of human genetic disorder also introduces new concepts in biology. One of the best examples given in this chapter is the idea of recessive oncogenes, first developed as a result of studying retinoblastoma. Another exciting area is derived from the cloning of *CFTR*, which has opened up new research directions in understanding chloride channels and membrane transports (Tsui 1991). Genes identified through the positional cloning approach also display unique features of the genome structure and organization. For example, the dystrophin gene with a size over 2 million base pairs is still the largest gene known in the mammalian genome, and most of the mutations in the gene are due to interstitial deletion; the NF1 gene contains at least three smaller genes embedded in one of its introns (White and O'Connell 1991); strong linkage disequilibrium is detected between DNA markers separated by hundreds of kilobases around the *CFTR* gene region; more than 100 different mutations have been identified within the *CFTR* gene, but no gross deletion has so far been detected within the 230-kb gene.

Finally, there has been tremendous competition in the field of disease gene cloning. Aside from trivial matters, such as personal ego, funding, and job security, there is actually a great deal of emotion associated with the research: It is difficult to describe the feelings of performing repeated searches for clone after clone without an open reading frame, constructing and screening genomic and cDNA libraries one after another, watching the come and go of a candidate gene, and working under the fear that the gene has perhaps been identified by another group. Nevertheless, although being the first to report appears to be the only consolation of this hard work, one should keep in mind that gene identification is in fact only the first step of true "reverse genetics" (i.e., study of function from the cloned gene) in understanding a disease, as vividly illustrated by the examples cited in this chapter.

Acknowledgments

The authors thank Francis Collins, Laura Bonetta, Annie Huang, Brenda Gallie, and Peter Ray for their careful reading of the manuscript and helpful comments. The research in the laboratory of L.-C.T. is supported by grants from the National Institutes of Health (DK-34944, DK-41980), the Canadian Cystic Fibrosis Foundation, the Cystic Fibrosis Foundation (USA), Medical Research Council (MRC) of Canada, Howard Hughes Medical Institute (HHMI), and the Canadian Network of Centres of Excellence. L.-C.T. is Sellers Chair in Cystic Fibrosis Research at the Hospital for Sick Children (Toronto) and also an MRC Scientist and HHMI International Scholar. The research in the laboratory of X.E. is supported by grants from the Fondo de Investigaciones Sanitrias de la Seguridad Social (90-E1254), the Dirección General de Investigación Científico y Técnica, the Comisión Interministerial de Ciencia y Tecnología, the Ministerio de Educación y Ciencia (24A), the Department d'Ensenyament de la Generalitat de Catalunya, and the Institut Català de la Salut.

References

Anderson, M.P., D.P. Rich, R.J. Gregory, A.E. Smith, and M.J. Welsh. 1991. Generation of cAMP-activated chloride currents by expression of CFTR.

Science **251:** 679.

Baehner, R.L., L.M. Kunkel, A.P. Monaco, J.L. Haines, P.M. Coneally, C. Palmer, N. Heerema, and S.H. Orkin. 1986. DNA linkage analysis of X chromosome-linked chronic granulomatous disease. *Proc. Natl. Acad. Sci.* **83:** 3398

Bakker, E., M.H. Hofker, N. Goor, J.L. Mandel, K. Wrogemann, K.E. Davies, L.M. Kunkel, H.F. Willard, W.A. Fenton, L. Sandkuyl, D. Majoor-Krakauer, A.J. Van Essen, M.G.J. Jahoda, E. Sachs, G.J.B. van Ommen, and P.L. Pearson. 1985. Prenatal diagnosis and carrier detection of Duchenne muscular dystrophy with closely linked RFLPs. *Lancet* **I:** 655.

Ballester, R., D. Marchuk, M. Boguski, A. Saulino, R. Letcher, M. Wigler, and F. Collins. 1990. The NF1 locus encodes a protein functionally related to mammalian GAP and yeast IRA proteins. *Cell* **63:** 851.

Barker, D., E. Wright, K. Nguyen, L. Cannon, P. Fain, D. Goldgar, D. Bishop, J. Carey, B. Baty, J. Kivlin, H. Willard, J.S. Waye, G. Greig, L. Leinwand, Y. Nakamura, P. O'Connell, M. Leppert, J.-M. Lalouel, R. White, and M. Skolnick. 1987. Gene for von Recklinghausen neurofibromatosis is in the pericentromeric region of chromosome 17. *Science* **236:** 1100.

Bates, G.P., M.E. MacDonlad, S. Baxendale, S. Youngman, C. Lin, W.L. Whaley, J.J. Wasmuth, J.F. Gusella, and H. Lehrach. 1991. Defined physical limits of the Huntington disease gene candidate region. *Am. J. Hum. Genet.* **49:** 7.

Beaudet, A., A. Bowcock, M. Buchwald, L. Cavalli-Sforza, M. Farrall, M.-C. King, K. Klinger, J.-M. Lalouel, M. Lathrop, S. Naylor, J. Ott, L.-C. Tsui, B. Wainwright, P. Watkins, R. White, and R. Williamson. 1986. Linkage of cystic fibrosis to two tightly linked DNA markers: Joint report from a collaborative study. *Am. J. Hum. Genet.* **39:** 681.

Berger, W., J. Hein, J. Gedschold, I. Bauer, A. Speer, M. Farrall, R. Williamson, and C. Coutelle. 1987. Crossovers in two German cystic fibrosis families determine probe order for MET, 7C22 and XV-2c/CS.7. *Hum. Genet.* **77:** 197.

Bickmore, W.S. Christie, V. van Heyningen, N.D. Hastie, and D.J. Porteous. 1988. Hitchhiking from HRAS1 to the WAGR locus with CMGT markers. *Nucleic Acids Res.* **16:** 51.

Bird, A. 1986. CpG-rich islands and the function of DNA methylation. *Nature* **321:** 209.

Boat, T.F., M.J. Welsh, and A.L. Beaudet. 1989. Cystic fibrosis. In *The metabolic basis of inherited disease*, 6th ed. (ed. C.R. Scriver et al.), p. 2649. McGraw-Hill, New York.

Bonetta, L., S.E. Kuehn, A. Huang, D.J. Law, L.M. Kalikin, M. Koi, A.E. Reeves, B.H. Brownstein, H. Yeger, B.R.G. Williams, and A.P. Feinberg. 1990. Wilms' tumor locus on 11p13 defined by multiple CpG island-associated transcripts. *Science* **250:** 994.

Botstein, D. 1990. 1989 Allen Award Address. *Am. J. Hum. Genet.* **47:** 887.

Botstein, D., R.L. White, M. Skolnick, and R.W. Davis. 1980. Construction of a genetic linkage map in man using restriction fragment length polymorphisms. *Am. J. Hum. Genet.* **32:** 314.

Buchberg, A.M., H.G. Bedigan, N.A. Jenkins, and N.G. Copeland. 1990. Evi-2. A common integration site involved in murine myeloid leukemogenesis. *Mol. Cell Biol.* **10:** 4658.

Buckler, A.J., D. Chang, S.L. Graw, J.D. Brook, D.A. Habor, P.A. Sharp, and D.E.

Housman. 1991. Exon amplification: A strategy to isolate mammalian genes based on RNA splicing. *Proc. Natl. Acad. Sci.* **88:** 4005.
Burke, D.R., G.F. Carle, and M.V. Olson. 1987. Cloning of large segments of DNA into yeast by means of artificial chromosome vectors. *Science* **236:** 806.
Burke, D.T. 1991. The role of yeast artificial chromosome clones in generating genome maps. *Curr. Opin. Genet. Dev.* **1:** 69.
Burmeister, M. and H. Lehrach. 1986. Long-range restriction map around the Duchenne muscular dystrophy gene. *Nature* **324:** 582.
Call, K.M., T. Glaser, C.Y. Ito, A.J. Buckler, J. Pelletier, D.A. Habor, E.A. Rose, A. Kral, H. Yeger, W.H. Lewis, C. Jones, and D.E. Housman. 1990. Isolation and characterization of a zinc-finger polypeptide gene at the human chromosome 11 Wilms' tumor locus. *Cell* **60:** 509.
Caskey, C.T. 1991. Genetics of disease (editorial overview). *Curr. Opin. Genet. Dev.* **1:** 3.
Cavenee, W., N. Hastie, and E. Stanbridge, eds. 1989. *Current communications in molecular biology: Recessive oncogenes and tumor suppression*, p. 93. Cold Spring Harbor Laboratory Press, Cold Spring Harbor, New York.
Cavenee, W., R. Leach, T. Mohandas, P. Pearson, and R. White. 1984. Isolation and regional localization of DNA segments revealing polymorphic loci from human chromosome 13. *Am. J. Hum. Genet.* **36:** 10.
Cavenee, W.K., T.P. Dryja, R.A. Phillips, W.F. Benedict, R. Godbout, B.L. Gallie, A.L. Murphree, L.C. Strong, and R.L. White. 1983. Expression of recessive alleles by chromosome mechanisms in retinoblastoma. *Nature* **305:** 779.
Cawthon, R., R. Weiss, G. Xu, D. Viskochil, M. Culver, J. Stevens, M. Robertson, D. Dunn, R. Gesteland, P. O'Connell, and R. White. 1990. A major segment of the neurofibromatosis type 1 gene: cDNA sequence, genomic structure, and point mutations. *Cell* **62:** 193.
Cawthon, R.M., L.B. Andersen, A.M. Buchberg, G. Fu, P. O'Connell, D. Viskochil, R.B. Weiss, M. Wallace, D.A. Marchuk, M. Culver, J. Stevens, N.A. Jenskins, N.G. Copeland, F.S. Collins, and R. White. 1991. cDNA sequence and genomic structure of *EVI2B*, a gene lying within an intron of the neurofibromatosis type 1 gene. *Genomics* **9:** 446.
Collins, F.S. and S.M. Weissman. 1984. Directional cloning of DNA fragments at large distance from an initial probe: A circularization method. *Proc. Natl. Acad. Sci.* **81:** 6912.
Compton, D.A., M.M. Weil, C. Jones, V.M. Riccardi, L.C. Strong, and G.F. Saunders. 1988. Long range physical map of the Wilms' tumor-aniridia region on human chromosome 11. *Cell* **55:** 827.
Compton, D.A., M.M. Weil, L. Bonetta, A. Huang, C. Jones, H. Yeger, B.R.G. Williams, L.C. Strong, and G.F. Saunders. 1990. Definition of the limits of the Wilms' tumor locus on human chromosome 11p13. *Genomics* **6:** 309.
Connolly, M.J., R.H. Payne, G. Johnson, B.L. Gallie, P.W. Alderdice, W.H. Marshall, and R.D. Lawton. 1983. Familial EsD-linked, retinoblastoma with reduced penetrance and variable expressivity. *Hum. Genet.* **65:** 122.
Cotton, R.G.H., N.R. Rodrigues, and R.D. Campbell. 1988. Reactivity of cytocine and thymine in single-base-pair mismatches with hydroxylamine and osmium tetroxide and its application to the study of mutations. *Proc. Natl. Acad. Sci.* **85:** 4397.
Cowell, J.K., R.B. Wadey, D.A. Haber, K.M. Call, D.E. Housman, and J. Pritchard. 1991. Structural rearrangement of the WT1 gene in Wilms'

tumor cells. *Oncogene* 6: 595.
Cox, D.R., C.A. Pritchard, E. Uglum, D. Casher, J. Kobori, and R.M. Meyers. 1989. Segregation of the Huntington's disease region of human chromosome 4 in a somatic cell hybrid. *Genomics* 4: 397.
Cox, T.K., B. Kerem, J. Rommens, M.C. Iannuzzi, M. Drumm, M., F.S. Collins, M. Dean, L.-C. Tsui, and A. Chakravarti. 1989. Mapping of the cystic fibrosis gene using putative ancestral recombinants. *Am. J. Hum. Genet.* 45: A136.
Cremers, F.P.M., D.J.R. van de Pol, L.P.M. van Kerkhoff, B. Wieringa, and H.-H. Ropers. 1990a. Cloning of a gene that is rearranged in patients with choroideremia. *Nature* 347: 674.
Cremers, F.P.M., F. Brunsmann, T.J.R. van de Pol, I.H. Pawlowitzki, K. Paulsen, B. Wieringa, and H.-H. Ropers. 1987. Deletion of the DXS165 locus in patients with classical choroideremia. *Clin. Genet.* 32: 421.
Cremers, F.P.M., T.J.R. van de Pol, P.J. Diergaarde, B. Wieringa, R.L. Nussbaum, M. Schwartz, and H.-H. Ropers. 1989a. Physical fine mapping of the choroideremia locus using Xq21 deletions associated with complex syndromes. *Genomics* 4: 41.
Cremers, F.P.M., F. Brunsmann, W. Berger, E.P.M. van Kerkhoff, T.J.R. van de Pol, B. Wieringa, I.H. Pawlowitzki, and H.-H. Ropers. 1990b. Cloning of the breakpoints of a deletion associated with choroideremia. *Hum. Genet.* 86: 61.
Cremers, F.P.M., T.J.R. van de Pol, B. Wieringa, F.S. Collins, E.-M. Sankila, V.M. Siu, W.F. Flintoff, F. Brunsmann, L.A.J. Blonden, and H.-H. Ropers. 1989b. Chromosome jumping from the DXS165 locus allows molecular characterization of four microdeletions and a de novo chromosome X/13 translocation associated with choroideremia. *Proc. Natl. Acad. Sci.* 86: 7510.
Cremers, F.P.M., E.-M. Sankila, F. Brunsmann, M. Jay, B. Jay, A. Wright, A.J.L.G. Pinckers, M. Schwartz, D.J.R. van de Pol, B. Wieringa, A. de la Chapelle, I.H. Pawlowitzki, and H.-H. Ropers. 1990c. Deletions in patients with classical choroideremia vary in size from 45 kb to several megabases. *Am. J. Hum. Genet.* 47: 622.
Crowe, F., W. Schull, and J. Neel. 1956. *A clinical, pathological and genetic study of multiple neurofibromatosis*. Charles C. Thomas, Springfield, Illinois.
Davies, K.E., P.L. Pearson, P.S. Harper, J.M. Murray, T. O'Brien, M. Safarazi, and R. Williamson. 1983. Linkage analysis of two cloned DNA sequences flanking the Duchenne muscular dystrophy locus on the short arm of the human X chromosome. *Nucleic Acids Res.* 11: 2303.
Davies, L.M., R. Stallard, G.H. Thomas, P. Couillin, C. Junien, N.J. Nowak, and T.B. Shows. 1988. Two anonymous DNA segments distinguishing the Wilms' tumor and aniridia loci. *Science* 241: 840.
Davisson, M.T., P.A. Lalley, J. Peters, D.P. Doolittle, A.L. Hillyard, and A.G. Searle. 1990. Report of the comparative subcommittee for human and mouse homologies. *Cytogenet. Cell Genet.* 55: 434.
Deaven, L.L., M.A. van Dilla, M.F. Bartholdi, A.V. Carrano, L.S. Cram, J.C. Fuscoe, J.W. Gray, C.E. Hildebrand, R.K. Moyzis, and J. Perlman. 1986. Construction of human chromosome-specific DNA libraries from flow-sorted chromosomes. *Cold Spring Harbor Symp. Quant. Biol.* 51: 159.
DeCaprio, J.A., J.W. Ludlow, J. Figge, J.Y. Shew, C.M. Huang, W.H. Lee, E. Mar-

silo, E. Paucha, and D.M. Livingston. 1988. SV40 large tumor antigen forms a specific complex with the product of the retinoblastoma susceptibility gene. *Cell* **54:** 275.

Dietz, H.C., G.R. Cutting, R.E. Pyeritz, C.L. Maslen, L.Y. Sakai, G.M. Corson, E.G. Puffenberger, A. Hamosh, E.J. Nanthakumar, S.M. Curristin, G. Stetten, D.A. Meyers, and C.A. Francomano. 1991. Marfan syndrome caused by a recurrent de novo missense mutation in the fibrillin gene. *Nature* **352:** 337.

Donis-Keller, H., P. Green, C. Helms, S. Cartinhour, B. Weiffenbach, K. Stephens, T.P. Keith, and 26 others. 1987. A genetic linkage map of the human genome. *Cell* **51:** 319.

Dorin, J. R., J.D. Inglis, and D.J. Porteous. 1989. Selection of precise chromosomal targeting of a dominant marker by homologous recombination. *Science* **243:** 1357.

Drayna, D. and R. White. 1985. The genetic linkage map of the human X chromosome. *Science* **230:** 753.

Drumm, M.L., C.L. Smith, M. Dean, J.L. Cole, M.C. Iannuzzi, and F.S. Collins. 1988. Physical mapping of the cystic fibrosis region by pulsed-field gel electrophoresis. *Genomics* **2:** 346.

Drumm, M.L., H.A. Pope, W.H. Cliff, J.M. Rommens, S.A. Marvin, L.-C. Tsui, F.S. Collins, R.A. Frizzell, and J.M. Wilson. 1990. Correction of the cystic fibrosis defect in vitro by retrovirus-mediated gene transfer. *Cell* **62:** 1227.

Dryja, T.P., J.M. Rapaport, J.M. Joyce, and R.A. Petersen. 1986. Molecular detection of deletions involving band q14 of chromosome 13 in retinoblastoma. *Proc. Natl. Acad. Sci.* **83:** 7391.

Dryja, T.P., T.L. MacGee, E. Reichel, L.B. Hahn, G.S. Cowley, D.W. Yandell, M.A. Snadberg, and E.L. Berson. 1990. A point mutation of the rhodopsin gene in one form of retinitis pigmentosa. *Nature* **343:** 364.

Dunn, J.M., R.A. Phillips, A.J. Becker, and B.L. Gallie. 1988. Identification of germline and somatic mutations affecting the retinoblastoma gene. *Science* **241:** 1797.

Duyk, G.M., S.W. Kim, R.M. Myers, and D.R. Cox. 1990. Exon trapping: A genetic screen to identify candidate transcribed sequences in cloned mammalian genomic DNA. *Proc. Natl. Acad. Sci.* **87:** 8995.

Dyson, N., P.M. Howley, K. Münger, and E. Harlow. 1989. The human papilloma virus-16 E7 oncoprotein is able to bond to the retinoblastoma gene product. *Science* **242:** 934.

Egan, C., T.N. Jelsma, J.A. Howe, S.T. Bayley, B. Ferguson, and P.E. Branton. 1989. Mapping of the cellular protein-binding sites on the products of early-region 1A of human adenovirus type 5. *Mol. Cell. Biol.* **8:** 3955.

Eiberg, H., J. Mohr, K. Schmiegelow, L.S. Nielsen, and R. Williamson. 1985. Linkage relationships of paraoxonase (PON) with other markers: Indication of PON-cystic fibrosis syntheny. *Clin. Genet.* **28:** 265.

Emery, A.E.H. 1987. *Duchenne muscular dystrophy.* Oxford University Press, United Kingdom.

Estivill, X., P.J. Scambler, B.J. Wainwright, K. Hawley, P. Frederick, M. Schwartz, M. Baiget, J. Kere, R. Williamson, and M. Farrall. 1987a. Patterns of polymorphism and linkage disequilibrium for cystic fibrosis. *Genomics* **1:** 257.

Estivill, X., M. Farrall, P.J. Scambler, G.M. Bell, K.M.F. Hawley, N.J. Lench, G.P.

Bates, H.C. Kruyer, P.A. Frederick, P. Stanier, E.K. Watson, R. Williamson, and B.J. Wainwright. 1987b. A candidate for the cystic fibrosis locus isolated by selection for methylation-free islands. *Nature* **326:** 840.

Farrall, M., B.J. Wainwright, G.L. Feldman, A. Beaudet, Z. Sretenovic, D. Halley, M. Simon, L. Dickerman, M. Devoto, G. Romeo, J.-C. Kaplan, A. Kitzis, and R. Williamson. 1988. Recombinations between IRP and cystic fibrosis. *Am. J. Hum. Genet.* **43:** 471.

Fearon, E.R., B. Vogelstein, and A.P. Feinberg. 1984. Somatic deletion and duplication of genes on chromosome 11 in Wilms' tumor. *Nature* **309:** 176.

Fountain, J., M. Wallace, M. Bruce, B. Seizinger, A. Menon, J. Gusella, V. Michels, M. Schmidt, G. Dewald, and F. Collins. 1989. Physical mapping of a translocation breakpoint in neurofibromatosis. *Science* **244:** 1085.

Francke, U. 1990. A gene for Wilms' tumor? *Nature* **343:** 692.

Francke, U., L.B. Holmes, L. Atkins, and V.M. Riccardi. 1979. Aniridia-Wilms' tumor association: Evidence for specific deletion of 11p13. *Cytogenet. Cell Genet.* **24:** 185.

Francke, U., H.D. Ochs, B. de Martinville, J. Giacalone, V. Lindgren, C. Distèche, R.A. Pagon, M.H. Hofker, G.J.B. van Ommen, P.L. Pearson, and R.J. Wedgwood. 1985. Minor Xp21 chromosome deletion in a male associated with expression of Duchenne muscular dystrophy, chronic granulomatous disease, retinitis pigmentosa, and McLeod syndrome. *Am. J. Hum. Genet.* **37:** 250.

Friend, S.H., R. Bernards, S. Rogelj, R.A. Weinberg, J.M. Rapaport, D.M. Albert, and T.P. Dryja. 1986. A human DNA segment with properties of the gene that predisposes to retinoblastoma and osteosarcoma. *Nature* **323:** 643.

Friend, S.H., J.M. Horowitz, M.R. Gerber, X.F. Wang, E. Boenmann, F.P. Li, and R.A. Weinberg. 1987. Deletions of a DNA sequence in retinoblastomas and mesenchymal tumors: Organization of the sequence and its encoded protein. *Proc. Natl. Acad. Sci.* **84:** 9059.

Fung, Y.K., A.L. Murphree, A. Tang, J. Qian, S.H. Hinrichs, and W.F. Benedict. 1987. Structural evidence for the authenticity of the human retinoblastoma gene. *Science* **236:** 1657.

Gallie, B.L., J.A. Squire, A. Goddard, J.M. Dunn, M. Canton, D. Hinton, X. Zhu, and R.A. Phillips. 1990. Mechanism of oncogenesis in retinoblastoma. *Lab. Invest.* **62:** 394.

Gessler, M. and G.A. Bruns. 1989. A physical map around the WAGR complex on the short arm of chromosome 11. *Genomics* **5:** 43.

Gessler, M., A. Poustka, W. Cavenee, R.L. Reve, S.H. Orkin, and G.A.P. Bruns. 1990. Homozygous deletion in Wilms' tumors of a zinc-finger gene identified by chromosome jumping. *Nature* **343:** 774.

Gessler, M., G.H. Thomas, P. Coullin, C. Junien, B.C. McGillivrary, M. Hayden, G. Jaschek, and G.A.P. Bruns. 1989. A deletion map of the WAGR region on chromosome 11. *Am. J. Hum. Genet.* **44:** 486.

Glaser, T., J. Lane, and D.E. Housman. 1990a. A mouse model of the aniridia-Wilms' tumor deletion syndrome. *Science* **259:** 994.

Glaser, T., E.A. Rose, H. Morse, D. Housman, and C. Jones. 1990b. A panel of irradiation-reduced hybrids selectively retaining human chromosome 11p13: Their structure and use to purify the WAGR gene complex. *Genomics* **6:** 48.

Goddard, A.D., H. Balakier, M. Canton, J. Dunn, J. Squire, E. Reyes, A. Becker,

R.A. Phillips, and B.L. Gallie. 1988. Infrequent genomic rearrangement and normal expression of the putative RB1 gene in retinoblastoma tumors. *Mol. Cell. Biol.* **8:** 2082.

Goedbloed, J. 1942. Mode of inheritance in choroideremia. *Ophthalmologica* **104:** 308.

Goldgar, D.E., P. Green, D.M. Parry, and J.J. Mulvihill. 1989. Multipoint linkage analysis in neurofibromatosis type 1: An international collaboration. *Am. J. Hum. Genet.* **44:** 6.

Goodfellow, P.N., K.E. Davies, and H.H. Ropers. 1985. Report of the committee on the genetic constitution of the X and Y chromosomes: Human gene mapping 8. *Cytogenet. Cell. Genet.* **40:** 296.

Goss, S.J. and H. Harris. 1975. New methods for mapping genes in human chromosomes. *Nature* **255:** 680.

Greenstein, R.M., M.P. Readon, and T.S. Chan. 1977. An X-autosome translocation in a girl with Duchenne muscular dystrophy, evidence for DMD gene localisation. *Pediatr. Res.* **11:** 475A.

Groden, J., A. Thliveris, W. Samowitz, M. Carlson, L. Gelbert, H. Albertsen, G. Joslyn, J. Stevens, L. Spirio, M. Robertson, L. Sargeant, K. Krapcho, E. Wolff, R. Burt, J.P. Hughes, J. Warrington, J. McPherson, J. Wasmuth, D. Le Paslier, H. Abderrahim, D. Cohen, M. Leppert, and R. White. 1991. Identification and characterization of the familial adenomatous polyposis coli gene. *Cell* **66:** 589.

Haber, D.A., A.J. Buckler, T. Glaser, K.M. Call, J. Pelletier, R.L. Sohn, E.C. Douglass, and D.E. Housman. 1990. An internal deletion within an 11p13 zinc finger gene contributes to the development of Wilms' tumor. *Cell* **61:** 1257.

Harbour, J.W., S.L. Lai, P.J. Whang, A.F. Gazdar, J.D. Minna, and F.J. Kaye. 1988. Abnormalities in structure and expression of the human retinoblastoma gene in SCLC. *Science* **241:** 353.

Harris, P.C., N.J. Barton, D.R. Higgs, S.T. Reeders, and A.O. Wilkie. 1990. A long range-restriction map between the alpha-globin complex and a marker closely linked to the polycystic kidney disease 1 (PKD1) locus. *Genomics* **7:** 195.

Hodgson, S.V., M.E. Robertson, C.N. Fear, J. Goodship, S. Malcolm, B. Jay, M. Bobrow, and M.E. Prembly. 1987. Prenatal diagnosis of X-linked choroideremia with mental retardation, associated with a cytologically detectable X-chromosome deletion. *Hum. Genet.* **75:** 286.

Hoffman, E.P., R.H. Brown, and L.M. Kunkel. 1987a. Dystrophin: The protein product of the Duchenne muscular dystrophy gene. *Cell* **51:** 919.

Hoffman, E.P., A.P. Monaco, C. Feener, and L.M. Kunkel. 1987b. Conservation of the Duchenne muscular dystrophy gene in mice and humans. *Science* **238:** 347.

Huang, A., C.E. Campbell, L. Bonetta, M.S. McAndrews-Hill, S. Chilton-MacNeil, M.J. Coppers, D.J. Law, A.P. Feinberg, H. Yeger, and B.R.G. Williams. 1990. Tissue, developmental, and tumor-specific expression of divergent transcripts in Wilms' tumor. *Science* **250:** 991.

Huang, H.Y., J.K. Lee, J.Y. Shew, P.L. Chen, R. Bookstein, T. Friedmann, E.Y. Lee, and W.H. Lee. 1988. Suppression of the neoplastic phenotype by replacement of the RB gene in human cancer cells. *Science* **242:** 1563.

Huff, V., H. Miwa, D.A. Haber, K.M. Call, D.E. Housman, L.C. Strong, and G.A.

Saunders. 1991. Evidence for WT1 as a Wilms' tumor (WT) gene: Intragenic germinal deletion in bilateral WT. *Am. J. Hum. Genet.* **48:** 997.
Jay, M., A.F. Wright, J.F. Clayton, M. Deans, M. Dempster, S.S. Bhattacharya, and B. Jay. 1986. A genetic linkage study of choroideremia. *Ophthalmic Paediatr. Genet.* **7:** 201.
Jobs, A., D. Klein-Bolting, A.S. Jandel, A. Driesel, K. Olek, and K.-H. Grzeschik. 1990. Regional assignment of 41 human DNA fragments on chromosome 7 by means of a somatic cell hybrid panel. *Hum. Genet.* **84:** 147.
Joslyn, G., M. Carlson, A. Thliveris, H. Albertsen, L. Gelbert, W. Samowitz, J. Groden, J. Stevens, L. Spirio, M. Robertson, L. Sargeant, K. Krapcho, E. Wolff, R. Burt, J.P. Hughes, J. Warrington, J. McPherson, J. Wasmuth, D. Le Paslier, H. Abderrahim, D. Cohen, M. Leppert, and R. White. 1991. Identification of deletion mutations and three new genes at the familial polyposis locus. *Cell* **66:** 601.
Junien, C., C. Turleau, J. de Grouchy, R. Said, M.-O. Rethore, R. Tenconi, and J.L. Dufier. 1980. Regional assignment of catalase (CAT) gene to band 11p13: Association with the aniridia-Wilms' tumor-gonadal blastoma (WAGR) complex. *Ann. Genet.* **23:** 165.
Kan, Y.W. and A.M. Dozy. 1978. Polymorphism of DNA sequence adjacent to the human β-globin structural gene: Relationship to sickle mutation. *Proc. Natl. Acad. Sci.* **75:** 5631.
Kartner, N., J.W. Hanrahan, T.J. Jensen, A.L. Naismith, S. Sun, C.A. Ackenley, E.F. Reyes, L.-C. Tsui, J.M. Rommens, C.E. Bear, and J.R. Riordan. 1991. Expression of the cystic fibrosis gene in non-epithelial invertebrate cells produces a regulated anion conductance. *Cell* **64:** 681.
Kenwrick, S., M. Patterson, A. Speer, K. Fischbeck, and K.E. Davies. 1987. Molecular analysis of the Duchenne muscular dystrophy region using pulsed field gel electrophoresis. *Cell* **48:** 351.
Kerem, B., J.M. Rommens, J.A. Buchanan, D. Markiewicz, T.K. Cox, A. Chakravarti, M. Buchwald, and L.-C. Tsui. 1989. Identification of the cystic fibrosis gene: Genetic analysis. *Science* **245:** 1073.
Kinzler, K.W., M.C. Nilbert, L.-K. Su, B. Vogelstein, T.M. Bryan, D.B. Levy, K.J. Smith, A.C. Preisinger, P. Hedge, D. McKechnie, R. Finniear, A. Markham, J. Groffen, M.S. Bogushi, S.F. Altschul, A. Horri, H. Ando, Y. Miyoshi, Y. Miki, I. Nishisho, and Y. Nakamura. 1991. Identification of FAP locus gene from chromosome 5q21. *Science* **253:** 661.
Knowlton, R.G., O. Cohen-Haguenauer, V.C. Nguyen, J. Frezal, V. Brown, D. Barker, J.C. Braman, J.W. Schumm, L.-C. Tsui, M. Buchwald, and H. Donis-Keller. 1985. A polymorphic DNA marker linked to cystic fibrosis is located on chromosome 7. *Nature* **318:** 380.
Knudson, A.G.J. 1971. Mutation and cancer: Statistical study of retinoblastoma. *Proc. Natl. Acad. Sci.* **68:** 820.
———. 1985. Hereditary cancer, oncogenes and antioncogenes. *Cancer Res.* **45:** 1437.
Koenig, M., A.P. Monaco, and L.M. Kunkel. 1988. The complete sequence of dystrophin predicts rod-shaped cytoskeletal protein. *Cell* **53:** 219.
Koenig, M., E.P. Hoffman, C.J. Bertelson, A.P. Monaco, C. Feener, and L.M. Kunkel. 1987. Complete cloning of the Duchenne muscular dystrophy (DMD) gene and preliminary genomic organization of the DMD gene in normal and affected individuals. *Cell* **50:** 509.

Koenig, M., A.H. Beggs, M. Moyer, S. Scherpf, K. Heindrich, T. Bettecken, G. Meng, C.R. Müller, M. Lindlöf, H. Kaariainen, A de la Chapelle, A. Kiuru, M.-L. Savontaus, H. Gilgenkrantz, D. Récan, J. Chelly, J.-C. Kaplan, A.E. Covone, N. Archidiacono, G. Romeo, S. Liechti-Gallati, V. Schneider, S. Braga, H. Moser, B.T. Darras, P. Murphy, U. Francke, J.S. Chen, G. Morgan, M. Denton, C.R. Greenberg, K. Wrogemann, L.A.J. Blonden, H.M.B. van Paasen, G.J.B. van Ommen, and L.M. Kunkel. 1989. The molecular basis for the Duchenne versus Becker muscular dystrophy: Correlation of severity with type of deletion. *Am. J. Hum. Genet.* 45: 498.

Koufos, J. M.F. Hansen, B.C. Lampkin, M.L. Workman, N.G. Copeland, N.A. Jenkins, and W. Cavenee. 1984. Loss of alleles at loci on human chromosome 11 during genesis of Wilms' tumor. *Nature* 309: 170.

Kunkel, L.M., A.P. Monaco, W. Middlesworth, H. Ochs, and S.A. Latt. 1985. Specific cloning of DNA fragments absent from the DNA of a male patient with an X-chromosome deletion. *Proc. Natl. Acad. Sci.* 82: 4778.

Kunkel, L.M. and co-authors. 1986. Analysis of deletions in DNA of patients with Becker and Duchenne muscular dystrophy. *Nature* 322: 73.

Lalande, M., T.P. Dryja, R.R. Schreck, J. Shipley, A. Flint, and S.A. Latt. 1984. Isolation of human chromosome 13-specific DNA sequences cloned from flow-sorted chromosomes and potentially linked to the retinoblastoma locus. *Cancer Genet. Cytogenet.* 13: 283.

La Spada, A.R., E.M. Wilson, D.B. Lubahn, A.E. Harding, and K.H. Fischbeck. 1991. Androgen receptor gene mutations in X-linked spinal and bulbar muscular atrophy. *Nature* 352: 77.

Ledbetter, D. H. and W. Cavenee. 1989. Wilms' tumor. In *The metabolic basis of inherited disease*, 6th ed. (ed. C.R. Scriver et al.), p. 343. McGraw-Hill, New York.

Ledbetter, D.H., D.C. Rich, P. O'Connell, M. Leppert, and J.C. Carey. 1989. Precise localization of NF1 to 17q11.2 by balanced translocation. *Am. J. Hum. Genet.* 44: 20.

Lee, B., M. Godfrey, E. Vitale, H. Hori, M.-G. Mattei, M. Sarfarazi, P. Tsipouras, F. Ramirez, and D.W. Hollister. 1991. Linkage of Marfan syndrome and a phenotypically related disorder to two different fibrillin genes. *Nature* 352: 330.

Lee, E.Y. and W.H. Lee. 1986. Molecular cloning of the human esterase D gene, a genetic marker of retinoblastoma. *Proc. Natl. Acad. Sci.* 83: 6337.

Lee, E.Y., H. To, J.Y. Shew, R. Bookstein, P. Scully, and W.H. Lee. 1988. Inactivation of the retinoblastoma susceptibility gene in human breast cancer. *Science* 241: 218.

Lee, W.H., R. Bookstein, F. Hong, L.J. Young, J.Y. Shew, and E.Y. Lee. 1987a. Human retinoblastoma susceptibility gene: Cloning, identification, and sequence. *Science* 235: 1394.

Lee, W.H., J.Y. Shew, F.D. Hong, T.W. Sery, L.A. Donoso, L.J. Young, R. Bookstein, and E.Y. Lee. 1987b. The retinoblastoma susceptibility gene encodes a nuclear phospho-protein associated with DNA binding activity. *Nature* 329: 642.

Lesko, J.G., R.A. Lewis, and R.L. Nussbaum. 1987. Multipoint linkage analysis of loci in the proximal long arm of the human X chromosome: Application to mapping the choroideremia locus. *Am. J. Hum. Genet.* 40: 303.

Lewis, R.A., R.L. Nussbaum, and R. Ferrell. 1985. Mapping X-linked ophthalmic diseases: Provisional assignment of the locus for choroideremia to Xq13-24. *Ophthalmology* **92**: 800.

Lewis, W.H., H. Yeger, L. Bonetta, H.S.L. Chan, J. Kang, C. Junien, J. Cowell, C. Jones, and L.A. Defoe. 1988. Homozygous deletion of a DNA marker from chromosome 11p13 in sporadic Wilms' tumor. *Genomics* **3**: 25.

Lüdecke, H.-J., G. Senger, U. Claussen, and B. Horsthemke. 1989. Cloning defined regions of the human genome by microdissection of banded chromosomes and enzymatic amplification. *Nature* **338**: 348.

Martin, G., D. Viskochil, G. Gollag, P. McCare, W. Crosier, H. Haubruck, L. Conroy, R. Clark, P. O'Connell, R. Cawthon, M.A. Innis, and F. McCormick. 1991. The GAP-related domain of the neurofibromatosis type 1 gene product interacts with rasp21. *Cell* **63**: 843.

Maslen, C.L., G.M. Corson, B.K. Maddox, R.W. Glanville, and L.Y. Sakai. 1991. Partial sequence of a candidate gene for Marfan syndrome. *Nature* **352**: 334.

McWilliam, P., G.J. Farrar, P. Kenna, D.G. Bradley, M.M. Humphries, E.M. Sharp, D.J. McConnell, M. Lawlor, D. Sheils, C. Ryan, K. Stevens, S.P. Daiger, and P. Humphries. 1989. Autosomal dominant retinitis pigmentosa (ADRP): Localization of an ADRP gene to the long arm of chromosome 3. *Genomics* **5**: 619.

Melmer, G., R. Sood, J. Rommens, D. Rego, L.-C. Tsui, and M. Buchwald. 1990. Isolation of clones on chromosome 7 that contain recognition sites for rare-cutting enzymes by oligonucleotide hybridization. *Genomics* **7**: 173.

Merry, D.E., J.G. Lesko, V. Siu, W.F. Flintoff, F. Collins, R.A. Lewis, and R.L. Nussbaum. 1990. DXS165 detects a translocation breakpoint in a woman with choroideremia and a de novo X;13 translocation. *Genomics* **6**: 609.

Mikol, D.D., J.R. Gulcher, and K. Stefansson. 1990. The oligodenocyte-myelin glycoprotein belongs to a distinct family of proteins and contains the HNK-1 carbohydrate. *J. Cell Biol.* **110**: 471.

Monaco, A.P. and L.N. Kunkel. 1988. Cloning of the Duchenne/Becker muscular dystrophy locus. *Adv. Hum. Genet.* **17**: 61.

Monaco, A.P., R.L. Neve, C. Colletti-Feener, C.L. Bertelson, D.M. Kunit, and L.M. Kunkel. 1986. Isolation of candidate cDNAs for portions of the Duchenne muscular dystrophy gene. *Nature* **323**: 646.

Monaco, A.P., C.J.H. Bertelson, W. Middlesworth, C.-A. Colletti, J. Aldridge, K.H. Fischbeck, R. Bartlett, M.A. Pericak-Vance, A.D. Roses, and L.M. Kunkel. 1985. Detection of deletions spanning the Duchenne muscular dystrophy locus using a tightly linked DNA segment. *Nature* **316**: 842.

Myers, R.M. and T. Maniatis. 1986. Recent advances in the development of methods for detecting single-base substitutions associated with human genetics diseases. *Cold Spring Harbor Symp. Quant. Biol.* **51**: 275.

Nakamura, Y., M. Leppert, P. O'Connell, R. Wolff, T. Holm, M. Culver, C. Martin, E. Fujimoto, M. Hoff, E. Kumlin, and R. White. 1987. Variable number of tandem repeat (VNTR) markers for human gene mapping. *Science* **235**: 1616.

Nathans, J., D. Thomas, and D.S. Hogness. 1986a. Molecular genetics of human color vision: The genes encoding blue, green and red pigments. *Science* **232**: 193.

Nathans, J., T.P. Piantanida, R.L. Eddy, T.B. Shows, and D.S. Hogness. 1986b. Molecular genetics and inherited variations in human color vision. *Science*

232: 203.

Nishisho, I., Y. Nakamura, Y. Miyoshi, Y. Miki, H. Ando, A. Horii, K. Koyama, J. Utsunomiya, S. Baba, P. Hedge, A. Markham, A.J. Krush, G. Petersen, S.R. Hamilton, M.C. Nilbert, D.B. Levy, T.M. Bryan, A.C. Preisinger, K.J. Smith, L.-K. Su, K.W. Kinzler, and B. Vogelstein. 1991. Mutations of chromosome 5q21 genes in FAP and colorectal cancer patients. *Science* **253**: 665.

Nussbaum, R.L., R.A. Lewis, J.G. Lesko, and R. Ferrell. 1985. Choroideremia is linked to the restriction fragment length polymorphism DXYS1 at Xq13-21. *Am. J. Hum. Genet.* **37**: 473.

Nussbaum, R.L., J.G. Lesko, R.A. Lewis, S.A. Ledbetter, and D.H. Ledbetter. 1987. Isolation of anonymous DNA sequences from within a submicroscopic X chromosome deletion in a patient with choroideremia, deafness, and mental retardation. *Proc. Natl. Acad. Sci.* **84**: 6521.

O'Connell, P., R. Leach, R. Cawthon, M. Culver, J. Stevens, D. Viskochil, R.E.K. Fournier, D. Rich, D.H. Ledbetter, and R. White. 1989a. Two von Recklinghausen neurofibromatosis translocations map within a 600-kb segment of 17q11.2. *Science* **244**: 1087.

O'Connell, P., D. Viskochil, A.M. Buchberg, J. Fountain, R. Cawthon, M. Culver, J. Stevens, J. Rich, D.H. Ledbetter, M. Wallace, J.C. Carey, N.A. Jenkins, N.G. Copeland, F. Collins, and R. White. 1990. The human homolog of murine Evi-2 lies between two translocation breakpoints associated with von Recklinghausen neurofibromatosi. *Genomics* **7**: 547.

O'Connell, P., R. Leach, D.H. Ledbetter, R. Cawthon, M. Culver, J.R. Eldrige, A.K. Frej, T.R. Holm, E. Wolff, M.J. Thayer, A.J. Schafer, J.W. Fountain, M.R. Wallace, F.S. Collins, M.H. Skolnick, D.C. Rich, R.E.K. Fournier, B.J. Baty, J.C. Carey, M.F. Leppert, G.M. Lathrop, J.-M. Lalouel, and R. White. 1989b. Fine structure DNA mapping studies of the chromosomal region harboring the genetic defect in neurofibromatosis type 1. *Am. J. Hum. Genet.* **44**: 51.

Orita, M., Y. Suzuki, T. Sekiya, and K. Hayashi. 1989. Rapid and sensitive detection of point mutations and DNA polymorphisms using the polymerase chain reaction. *Genomics* **5**: 874.

Orkin, S.H. 1986. Reverse genetics and human disease. *Cell* **47**: 845.

———. 1987. X-linked chronic granulomatous disease: From chromosomal position to the in vivo gene product. *Trends Genet.* **3**: 149.

Orkin, S.H., D.S. Goldman, and S.E. Sallan. 1984. Development of homozygosity for chromosome 11p markers in Wilms' tumor. *Nature* **309**: 172.

Paulsen, K., S. Forrest, G. Scherer, H.-H. Ropers, and K. Davies. 1986. Regional localization of X chromosome short arm probes. *Hum. Genet.* **74**: 155.

Porteous, D.J., W. Bickmore, S. Christie, P.A. Byod, G. Granston, J.M. Flether, J.R. Gosden, D. Rout, A. Seawright, K.O.J. Simola, V. van Heyningen, and N.D. Hastie. 1987. HRAS1 selected chromosome transfer generates markers that colocalize aniridia- and genitourinary dysplasia-associated translocation breakpoints and the Wilms' tumor gene within 11p13. *Proc. Natl. Acad. Sci.* **84**: 5355.

Poustka, A. and H. Lehrach. 1986. Jumping libraries and linking libraries: The next generation of molecular tools in mammalian genetics. *Trends Genet.* **2**: 174.

Poustka, A., H. Lehrach, R. Williamson, and G. Bates. 1988. A long-range restriction map encompassing the cystic fibrosis locus and its closely linked ge-

netic markers. *Genomics* **2**: 337.
Pritchard, C., D.R. Cox, and R.M. Meyers. 1991. Invited editorial: The end is in sight for Huntington disease? *Am. J. Hum. Genet.* **49**: 1.
Ray, P.N., B. Belfall, C. Duff, C. Logan, V. Kean, M.W. Thompson, J.E. Sylvester, J.L. Gorski, R.D. Schmickel, and R.G. Worton. 1985. Cloning of the breakpoint of an X;21 translocation associated with Duchenne muscular dystrophy. *Nature* **318**: 672.
Reeves, A.E., P.J. Housiaux, R.J.M. Gardner, W.E. Chewing, R.M. Grindley, and L.J. Millow. 1984. Loss of a Harvey ras allele in sporadic Wilms' tumor. *Nature* **309**: 174.
Riccardi, V. and J. Eichner. 1986. *Neurofibromatosis: Phenotype, natural history, and pathogenesis.* Johns Hopkins University Press, Baltimore.
Riccardi, V.M. and R.A. Lewis. 1988. Penetrance of von Recklinghausen neurofibromatosis: A distinction between predecessors and descendants. *Am. J. Hum. Genet.* **42**: 284.
Riccardi, V.M., E. Sujansky, A.C. Smith, and U. Francke. 1978. Chromosome imbalance in the aniridia-Wilms' tumor association: 11p interstitial deletion. *Pediatrics* **61**: 604.
Rich, D.P., R.J. Gregory, M.P. Anderson, P. Manavalan, A.E. Smith, and M.J. Welsh. 1991. Effect of deleting the R domain on CFTR-generated chloride channels. *Science* **253**: 205.
Rich, D.P., M.P. Anderson, R.J. Gregory, S.H. Cheng, S. Paul, D.M. Jefferson, J.D.M. McCann, K.W. Klinger, A.E. Smith, and M.J. Welsh. 1990. Expression of cystic fibrosis transmembrane conductance regulator corrects defective chloride channel regulation in cystic fibrosis airway epithelial cells. *Nature* **347**: 358.
Riordan, J.R., J.M. Rommens, B. Kerem, N. Alon, R. Rozmahel, Z. Grzelchak, J. Zielenski, S. Lok, N. Plavsic, J.-L. Chou, M.L. Drumm, M.C. Iannuzzi, F.S. Collin, and L.-C. Tsui. 1989. Identification of the cystic fibrosis gene: Cloning and characterization of complementary DNA. *Science* **245**: 1066.
Röhme, D., H. Fox, B. Herrmann, A.-M. Frischauf, J.-E. Edström, P. Mains, L.M. Silver, and H. Lehrach. 1984. Molecular clones of the mouse t complex derived from microdissected metaphase chromosomes. *Cell* **36**: 783.
Rommens, J.M., S. Zengerling-Lentes, B. Kerem, G. Melmer, M. Buchwald, and L.-C. Tsui. 1989a. Physical localization of two DNA markers closely linked to the cystic fibrosis by pulsed field gel electrophoresis. *Am. J. Hum. Genet.* **45**: 932.
Rommens, J.M., S. Dho, C.E. Bear, N. Kartner, D. Kennedy, J.R. Riordan, L.-C. Tsui, and J.K. Foskett. 1991. Cyclic-AMP-inducible chloride conductance in mouse fibroblast lines stably expressing human cystic fibrosis transmembrane conductance regulator (CFTR). *Proc. Natl. Acad. Sci.* **88**: 7500.
Rommens, J.M., S. Zengerling, J. Burns, G. Melmer, B. Kerem, N. Plavsic, M. Zsiga, D. Kennedy, D. Markiewicz, R. Rozmahel, J.R. Riordan, M. Buchwald, and L.-C. Tsui. 1988. Identification and regional localization of DNA markers on chromosome 7 for the cloning of the cystic fibrosis gene. *Am. J. Hum. Genet.* **43**: 645.
Rommens, J.M., M.C. Iannuzzi, B. Kerem, M.L. Drumm, G. Melmer, M. Dean, R. Rozmahel, J.L. Cole, D. Kennedy, N. Hidaka, M. Zsiga, M. Buchwald, J.R. Riordan, L.-C. Tsui, and F.S. Collins. 1989b. Identification of the cystic fibrosis gene: Chromosome walking and jumping. *Science* **245**: 1059.

Rose, E.A., T. Glaser, C. Jones, C.L. Smith, W.H. Lewis, K.M. Call, M. Minden, E. Champagne, L. Bonetta, H. Yeger, and D.E. Housman. 1990. Complete physical map of the WAGR region of 11p13 localizes a candidate Wilms' tumor gene. *Cell* **60**: 495.

Royer-Pokora, B., L.M. Kunkel, A.P. Monaco, S.C. Goff, P.E. Newburger, R.L. Baehner, F.S. Cole, J.T. Curnutte, and S.H. Orkin. 1986. Cloning the gene for an inherited human disorder—chronic granulomatous disease—on the basis of its chromosomal location. *Nature* **322**: 32.

Sankila, E.M., A. del la Chapelle, J. Karna, H. Forsius, R. Frants, and A. Eriksson. 1987. Choroideremia: Close linkage to DXYS1 and DXYS12 demonstrated by segregation analysis and historical genealogical evidence. *Clin. Genet.* **31**: 315.

Scambler, P.J., H.-Y. Law, R. Williamson, and C.S. Cooper. 1986a. Chromosome mediated gene transfer of six DNA markers linked to the cystic fibrosis locus on human chromosome seven. *Nucleic Acids Res.* **14**: 7159.

Scambler, P.J., B.J. Wainwright, E. Watson, G. Bates, G. Bell, R. Williamson, and M. Farrall. 1986b. Isolation of a further anonymous informative DNA sequence from chromosome 7 closely linked to cystic fibrosis. *Nucleic Acids Res.* **14**: 1951.

Schmidt, M.A., V.V. Michels, and G.W. Dewald. 1987. Cases of neurofibromatosis with rearrangements of chromosome 17 involving band 111.2. *Am. J. Hum. Genet.* **28**: 771.

Schwartz, D.C. and C.R. Cantor. 1984. Separation of yeast chromosome-sized DNA by pulsed field gradient gel electrophoresis. *Cell* **37**: 67.

Schwartz, M., T. Rosenberg, E. Niebuhr, C. Lundsteen, H. Sardemann, O. Andersen, H.-M. Yang, and L.U. Lamm. 1986. Choroideremia: Further evidence for assignment of the locus to Xq13-Xq21. *Hum. Genet.* **74**: 449.

Segal, A.W., A.R. Cross, R.C. Garcia, N. Borregaard, N.H. Valerius, J.F. Soothill, and O.T.G. Jones. 1983. Absence of cytochrome b-245 in chronic granulomatous disease: A multicenter European evaluation of its incidence and relevance. *N. Engl. J. Med.* **308**: 245.

Seizinger, B., G. Rouleau, L. Ozelius, and co-authors. 1987. Genetic linkage of von Recklinghausen neurofibromatosis to the nerve growth factor receptor gene. *Cell* **49**: 589.

Shaw, D.J., H.G. Harley, J.D. Brook, and T.W. McKeithan. 1989. Long range restriction map of a region of human chromosome 19 containing the apolipoprotein gene, a CLL-associated translocation breakpoint and 2 polymorphic MluI site. *Hum. Genet.* **83**: 71.

Sinclair, A.H., P. Berta, M.S. Palmer, J.R. Hawkins, B.L. Griffiths, M.J. Smith, J.W. Foster, A.M. Frischauf, R. Lowell-Badge, and P.N. Goodfellow. 1990. A gene from the human sex-determining region encodes a protein with homology to a conserved DNA-binding motif. *Nature* **346**: 240.

Siu, V.M., J.R. Gonder, J.H. Jung, F.R. Sergovich, and W.F. Flintoff. 1990. Choroideremia associated with an X-autosomal translocation. *Hum. Genet.* **84**: 459.

Smeets, H., L. Bachinski, M. Coewinkel, J. Schepens, J. Hoeijmakers, M. van Duin, K.-H. Grzeschik, C.A. Weber, P. de Jong, M.J. Siciliano, and B. Wieringa. 1990. A long range restriction map of the human chromosome 19q13 region: Close physical linkage between CKMM and the ERCC1 and ERCC2 genes. *Am. J. Hum. Genet.* **46**: 492.

Smith, C.L. and C.R. Cantor. 1986. Approaches to physical mapping of the human genome. *Cold Spring Harbor Symp. Quant. Biol.* **51**: 115.

Sparkes, R.S., I. Klisak, D. Kaufman, T. Mohandas, A.J. Tobin, and J.F. McGinnis. 1986. Assignment of the rhodopsin gene to human chromosome three, region 3q21-3q24 by in situ hybridization studies. *Curr. Eye Res.* **5**: 797.

Sparkes, R.S., A.L. Murphree, R.W. Lingua, M.C. Sparkes, L.L. Field, S.J. Funderburk, and W.F. Benedict. 1983. Gene for hereditary retinoblastoma assigned to human chromosome 13 by linkage to esterase D. *Science* **219**: 971.

Sparkes, R.S., M.C. Sparkes, M.G. Wilson, J.W. Towner, W. Benedict, A.L. Murphree, and J.J. Yunis. 1980. Regional assignment of genes for human esterase D and retinoblastoma to chromosome band 13q14. *Science* **208**: 1042.

Squire, J., T.P. Dryja, J. Dunn, A. Goddard, T. Hoffman, M. Musarella, H.F. Willard, A.J. Becker, B.L. Gallie, and R.A. Phillips. 1986. Cloning of the esterase D gene: A polymorphic gene probe closely linked to the retinoblastoma locus on chromosome 13. *Proc. Natl. Acad. Sci.* **83**: 6573.

Tsui, L.-C. 1991. Probing the function of cystic fibrosis transmembrane conductance regulator. *Curr. Opin. Genet. Dev.* **1**: 4.

Tsui, L.-C. and M. Buchwald. 1991. Biochemical and molecular genetics of cystic fibrosis. *Adv. Hum. Genet.* **20**: 1530.

Tsui, L.-C., K. Buetow, and M. Buchwald. 1986. Genetic analysis of cystic fibrosis using linked DNA markers. *Am. J. Hum. Genet.* **39**: 720.

Tsui, L.-C., M. Buchwald, D. Barker, J.C. Braman, R.G. Knowlton, J. Schumm, H. Eiberg, J. Mohr, D. Kennedy, N. Plavsic, M. Zsiga, D. Markiewicz, G. Akots, V. Brown, C. Helms, T. Gravius, C. Parker, K. Rediker, and H. Donis-Keller. 1985. Cystic fibrosis locus defined by a genetically linked polymorphic DNA marker. *Science* **230**: 1054.

Turleau, C. and J. De Grouchy. 1987. Constitutional karyotypes in retinoblastoma. *Ophthalmic Paediatr. Genet.* **8**: 11.

van de Pol, T.J.R., F.P.M. Cremers, R.M. Brohet, B. Wieringa, and H.-H. Ropers. 1990. Derivation of clones from the choroideremia locus by preparative field inversion gel electrophoresis. *Nucleic Acids Res.* **18**: 725.

van Ommen, G.J.B., J.M.H. Verkerk, M.H. Hofker, A.P. Monaco, L.M. Kunkel, P. Ray, R. Worton, B. Wieringa, E. Bakker, and P.L. Pearson. 1986. A physical map of 4 million bp around the Duchenne muscular dystrophy gene on the human X chromosome. *Cell* **47**: 499.

Verellen-Dumoulin, C., M. Freund, R. de Meyer, C. Laterre, J. Frederic, M.W. Thompson, V.,D. Markovic, and R.G. Worton. 1984. Expression of an X-linked muscular dystrophy in a female due to translocation involving Xp21 and non-random X-inactivation. *Hum. Genet.* **67**: 115.

Verkerk, A.J.M.H., M. Pieretti, J.S. Sutcliffe, Y.-H. Fu, D.P.A. Kuhl, A. Pizzuti, O. Reiner, S. Richards, M.F. Victoria, F. Zhang, B.E. Eussen, G.-J.B. van Ommen, L.A.J. Blonden, G.J. Riggins, J.L. Chastain, C.B. Kunst, H. Galjaard, C.T. Caskey, D.L. Nelson, B.A. Oostra, and S.T. Warren. 1991. Identification of a gene (*FMR-1*) containing a CGG repeat coincident with a breakpoint cluster region exhibiting length variation in fragile X syndrome. *Cell* **65**: 905.

Viskochil, D., A. Buchberg, G. Xu, R. Cawthon, J. Stevens, R. Wolff, M. Culver, J.

Carey, N. Copeland, N. Jenkins, R. White, and P. O'Connell. 1990. Deletions and a translocation interrupt a cloned gene at the neurofibromatosis type 1 locus. *Cell* **62:** 187.
Vortkamp, A., M. Gessler, and K.-H. Grzschik. 1991. GL13 zinc-finger gene interrupted by translocations in Greig syndrome families. *Nature* **352:** 539.
Waardenburg, P.J. 1942. Choroideremie als Erbmerkmal. *Acta Ophthalmol.* **20:** 235.
Wainwright, B.J., P.J. Scambler, J. Schmidtke, E.A. Watson, H.-Y. Law, M. Farrall, H.J. Cooke, H. Eiberg, and R. Williamson. 1985. Localization of cystic fibrosis locus to human chromosome 7cen-q22. *Nature* **318:** 384.
Wainwright, B.J., P.J. Scambler, P. Stanier, E.K. Watson, G. Bell, C. Wicking, X. Estivill, M. Courtney, A. Boue, P.S. Pederson, R. Williamson, and M. Farrall. 1988. Isolation of a human gene with protein sequence similar to human and murine int-1 and the *Drosophila* segment polarity mutant wingless. *EMBO J.* **7:** 1743.
Wallace, M., D. Marchuk, L.B. Andersenm, and F.S. Collins. 1990a. Type 1 neurofibromatosis gene: Correction. *Science* **250:** 1749.
Wallace, M., D. Marchuk, L. Andersenm, R. Letcher, H. Odeh, A. Saulino, J. Fountain, A. Brereton, J. Nicholson, A. Mitchell, B. Brownstein, and F. Collins. 1990b. Type 1 neurofibromatosis gene: Identification of a large transcript disrupted in three NF1 patients. *Science* **249:** 181.
Weber, J.L. 1990. Informativeness of human $(dC-dA)_n \cdot (dG-dT)_n$ polymorphisms. *Genomics* **7:** 524.
White, R. and P. O'Connell. 1991. Identification and characterization of the gene for neurofibromatosis type 1. *Curr. Opin. Genet. Dev.* **1:** 15.
White, R., S. Woodward, Y. Nakamura, M. Leppert, P. O'Connell, M. Hoff, J. Herbst, J.-M. Lalouel, M. Dean, and G. Vande Woude. 1985. A closely linked genetic marker for cystic fibrosis. *Nature* **318:** 382.
White, R., M. Leppert, P. O'Connell, Y. Nakamura, S. Woodward, M. Hoff, J. Herbst, M. Dean, G.M. Vande Woude, M. Lathrop, and J.-M. Lalouel. 1986. Further linkage data on cystic fibrosis: The Utah study. *Am. J. Hum. Genet.* **41:** 944.
Whyte, P., N.M. Williamson, and E. Harlow. 1989. Cellular targets for transformation by the adenovirus E1A protein. *Cell* **56:** 67.
Wong, C., C.E. Dowling, R.K. Saiki, R.G. Higuchi, H.A. Erlich, and H.H. Kazazian, Jr. 1987. Characterization of beta-thalassemia mutation using direct genomic sequencing of amplified single copy DNA. *Nature* **330:** 384.
Worton, R.G. and M.W. Thompson. 1988. Genetics of Duchenne muscular dystrophy. *Annu. Rev. Genet.* **22:** 601.
Woychik, R.P., T.A. Stewart, L.G. Davies, P. D'Eustachio, and P. Leder. 1985. An inherited limb deformity created by insertional mutagenesis in transgenic mouse. *Nature* **318:** 36.
Wright, A.F., R.L. Nussbaum, S.S. Bhattacharya, M. Jay, J.G. Lesko, H.J. Evans, and B. Jay. 1990. Linkage studies and deletion screening in choroideremia. *J. Med. Genet.* **27:** 496.
Xu, G., P. O'Connell, D. Viskochil, R. Cawthon, M. Robertson, M. Culver, D. Dunn, J. Stevens, R. Gesteland, R. White, and R. Weiss. 1990. The neurofibromatosis type 1 gene encodes a protein related to GAP. *Cell* **62:** 599.
Yandell, D.W., T.A. Campbell, S.H. Dayton, R. Petersen, D. Walton, J.B. Little, A.

McConkie-Rosell, E.G. Buckley, and T.P. Dryja. 1989. Oncogenic point mutations in the human retinoblastoma gene: Their application to genetic counseling. *N. Engl. J. Med.* **321:** 1689.

Yokota, J., T. Akiyama, Y.K.T. Fung, W.F. Bennedict, Y. Namba, M. Hanaoka, M. Wada, T. Terasaki, Y. Shimosato, T. Sugimura, and M. Tarada. 1988. Altered expression of the retinoblastoma (RB) gene in small-cell lung carcinoma of the lung. *Oncogene* **3:** 471.

Yunis, J.J. and N. Ramsay. 1978. Retinoblastoma and subband deletion of chromosome 13. *Am. J. Dis. Child.* **132:** 161.

Zatz, M., A.M. Vianna-Morgante, P. Campos, and A.J. Diament. 1981. Translocation (X;6) in a female with Duchenne muscular dystrophy, implications for the localization of the DMD locus. *J. Med. Genet.* **18:** 442.

Zielenski, J., R. Rozmahel, D. Bozon, B. Kerem, Z. Grzelczak, J.R. Riordan, J.M. Rommens, and L.-C. Tsui. 1991. Genomic DNA sequence of the cystic fibrosis transmembrane conductance regulator (CFTR) gene. *Genomics* **10:** 214.

The Mouse *t* Complex Responder Locus

Linda C. Snyder and Lee M. Silver

Department of Molecular Biology
Princeton University
Princeton, New Jersey 08544-1014

During the last decade, advances in molecular technology have provided human geneticists with the ability to identify and characterize the genes that underlie a variety of disease phenotypes. Barely a week passes without an announcement that the genetic basis for another disease has been uncovered. The extraordinary rate at which advances in this area have been made is due almost entirely to the newfound ability of researchers to perform classic recombinational studies based on human pedigrees. The limitations to human geneticists in the past have been overcome with the development of large panels of polymorphic molecular markers that can be used to search for linkage with disease loci. With the use of modern methods for genomic analysis, linkage information can often be rapidly transformed into the identification of "candidate genes" that are likely to represent the disease locus under analysis.

Although candidates for many human disease genes have been uncovered, this represents only the first step toward the ultimate goal of understanding the molecular mechanisms responsible for each disease phenotype and how the phenotype might be overcome. An important tool in the pathway from candidate gene to molecular function is the ability to manipulate genes in defined ways so that an effect on phenotype can be observed; this is the classic approach of experimental genetics. However, for many complex diseases, phenotype can only be observed in the context of a living human being. This is a serious limitation since *Homo sapiens* is not an experimental organism and directed manipulation of the human genome cannot be considered. Fortunately, in most cases, it is possible to overcome this roadblock by moving to study the same gene in the mouse, a lesser cousin with the providential ability to mimic us in much of its biology.

Specific topics to be discussed include:

❏ the rationale for using the mouse as a tool for genetic analysis

❏ the history of investigations into mouse *t* haplotypes and their role as a model genetic system

❏ the genetics and physiology of the transmission ratio distortion phenotype expressed by male mice that carry a *t* haplotype

❏ the cloning and characterization of a strong candidate for the *t* complex responder locus, which is centrally involved in the transmission ratio distortion phenotype

❏ a model to explain the molecular basis for transmission ratio distortion

❏ final thoughts and future directions

GENETIC ANALYSIS IN THE MOUSE

Along with the development of new tools for recombinational analysis in mammals, there has been a rapid increase in the sophistication with which it is possible to manipulate the mouse genome. Two approaches in particular provide powerful tools for extended genetic analysis. The first is the construction of *transgenic mice*. The transgenic technology allows researchers to insert DNA from any source directly into the mouse germ line (Hogan et al. 1986). Unlimited numbers of each transgenic line can be bred for the study of gene expression and the effects of the *transgene* on phenotype. Whereas the transgenic approach allows investigators to add genetic material to mouse genomes, the *gene replacement* approach allows workers to alter genes already present in predefined ways (Capecchi 1989). This approach is based on the ability to convert cells from early mouse embryos into tissue culture lines, where genomes can be manipulated in a highly controlled and selective manner before they are returned to the embryo proper. Through this approach, lines of mice with altered genomes can be obtained for analysis of expression and mutant phenotypes.

Although the mouse provides an excellent model system for the analysis of genes first described in humans, it is often the case that interesting mutant phenotypes are observed in mice that have yet to be identified in humans. The rationale for pursuing an analysis of these phenotypes and their corresponding genes is the likelihood of unanticipated insight into human biology, since every mouse gene studied to date has a human homolog. In this paper, we describe an unusual gene system that has evolved the ability to alter drastically the differentiation and function of sperm in the mouse, which results in a phenotype known as transmission ratio distortion (TRD). The gene with a central role in TRD, the *t* complex responder (*Tcr*) locus, has been cloned and characterized. Although a similar phenotype has not been observed in humans, the *Tcr* gene is conserved and expressed in the human testes. Analysis of *Tcr* in particular, and TRD in general, may provide insight into aspects of male germ cell differentiation that are shared by mice and men. Here, we provide a brief review of the mouse *t* complex and the TRD phenotype, followed by a description of the approaches used to identify and characterize the *Tcr* locus at the molecular level.

THE *t* COMPLEX

The term *t* complex refers to a 20-cM region of DNA that corresponds approximately to the proximal half of mouse chromosome 17 (Fig. 1A). This region is uniquely defined only within the context of a naturally occurring, variant form known as a *t* haplotype, which is present in some 10–30% of mice in wild populations. Without the existence of variant *t* haplotypes, the wild-type *t* complex region would *not* be recognized as an entity distinct from surrounding genomic regions.

Wild mice bearing *t* haplotypes are not physically distinguishable from those not bearing them, so the discovery of this variant chromosome was serendipitous. In 1927, Dobrovolskaia-Zavadskaia was having difficulty maintaining a stock of mutant Brachyury mice in her laboratory. Brachyury is a dominant mutation at the *T* (for Tail) locus on chromosome 17, and heterozygous *T*/+ animals have a shortened, kinked tail (Dobrovolskaia-Zavadskaia 1927). To invigorate her Brachyury stock, Dobrovolskaia-Zavadskaia captured wild mice for use as mating partners in her colony. To her surprise, she found that a large percentage of the pups born in some matings had no tails at all (Dobrovolskaia-Zavadskaia and Kobozieff 1932). She concluded that some of the wild mice carried a recessive allele of *T*—"a *t* allele"—that was silent in the presence of the wild-type allele but interacted with *T* to produce a tailless phenotype in *T/t* individuals.

Soon after their discovery, *t* alleles were found to express a number of unrelated phenotypes. First, most, but not all, carry recessive lethal mutations that result in the death and resorption of homozygous embryos. However, *t* haplotypes isolated from different populations can carry different, complementing mutations such that t^x/t^y mice are often viable. To date, 16 complementation groups that act across the spectrum of prenatal development have been identified (Klein et al. 1984). A sec-

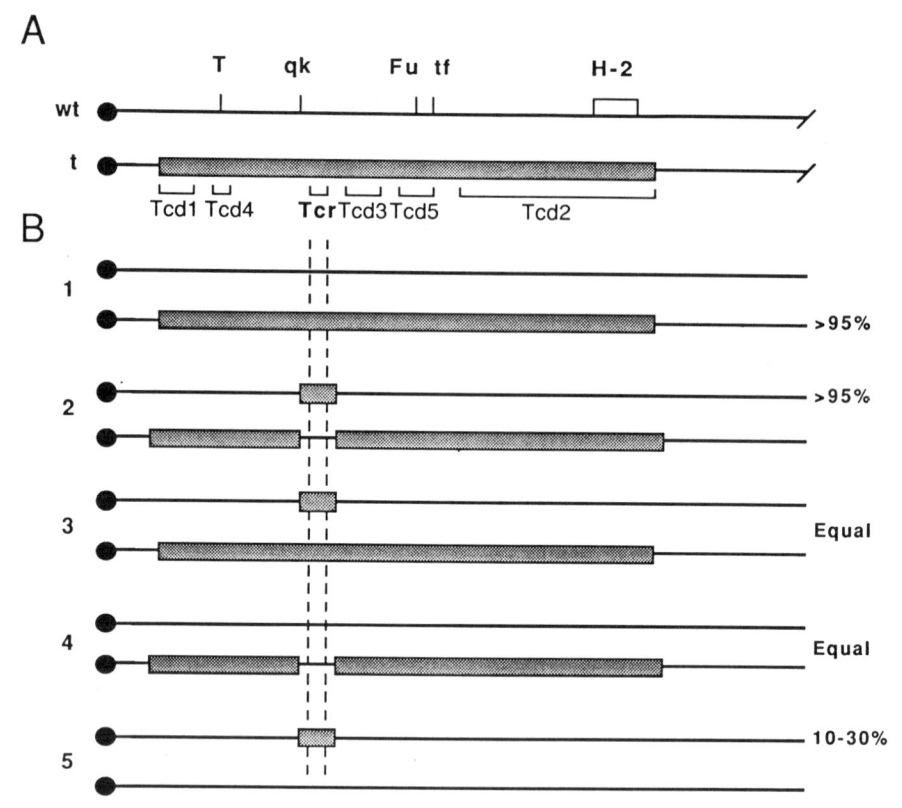

Figure 1 Maps of mouse chromosome 17. (A) Schematic diagram of wild-type (wt) and t forms of chromosome 17. The region of chromosome 17 defined by complete t haplotypes is shown as a shaded box, and wild-type DNA is denoted by a line. Anchor loci within this region of chromosome 17 include Brachyury (T), quaking (qk), Fused (Fu), and tufted (tf). Approximate locations of t alleles for the t complex responder (Tcr) and t complex distorter loci (Tcd-1–Tcd-5) are shown. (B) Genotypes of male mice that bear different combinations of wild-type and t haplotype DNA and corresponding levels of TRD. All loci are assumed to have wild-type and t alleles. (1) From a +/t male mouse, the t chromosome is transmitted to more than 95% of offspring. (2) TRD occurs if the Tcd^t alleles are present on the opposite chromosome as Tcr^t, but it is the Tcr^t-bearing chromosome that is transmitted at a high frequency. (3,4) The Tcr^t allele must be present and heterozygous for TRD to occur. (5) If the Tcr^t allele is present without Tcd^t alleles, the Tcr^t chromosome is transmitted at a low frequency.

ond observation was that t haplotypes cause a suppression of recombination along the proximal portion of chromosome 17 in all heterozygous animals. Another observation concerned the dramatic effect of t haplotypes on male fertility. Males born with two t haplotypes (t/t) are unconditionally sterile, whereas their female counterparts are fertile. Finally, males heterozygous for any complete t haplotype can transmit their t-bearing chromosome to 95% or more of their offspring in a clear departure from the laws of Mendel. This TRD phenotype is the critical defining feature that characterizes the t haplotype system.

The possibility that a single locus could have such diverse effects on development, fertility, and chromosome mechanics was truly extraordinary, and it led to the hypothesis that the T locus was the "master regulator" of all developmental processes. In particular, the finding that different t alleles could affect embryogenesis at such different points in time and space seemed to bolster this argument. The underpinning for the master gene hypothesis was swept away with the realization that so-called t alleles were not alterations at a single locus, but rather, they were alterations extending across what we now know to be half a chromosome (Lyon and Mason 1977). Thus, t alleles came to be known as t haplotypes to indicate their incorporation of many genetic elements. The different lethal complementation groups represent unrelated mutations at different loci within the region of recombination suppression (Artzt et al. 1982a), and evolutionary considerations now provide an explanation for the random accumulation of such mutations (Silver 1985). The effects of t haplotypes on sperm function and tail phenotype result from variant alleles at other unrelated loci (Lyon 1984). Finally, the finding of a series of nonoverlapping inversions that distinguish t haplotypes from their wild-type counterparts provides an explanation for both recombination suppression and the apparent inheritance of t haplotypes as single genetic entities (Artzt et al. 1982b; Herrmann et al. 1986; Sarvetnick et al. 1986; Hammer et al. 1989).

With this new understanding of t haplotype genetics came an understanding of their origin and the reason for their persistence in nearly all wild populations of mice (Silver 1985; Hammer et al. 1989). In essence, t haplotypes are selfish forms of the proximal half of chromosome 17 that have evolved the ability to transmit themselves from heterozygous $+/t$ males at the expense of the wild-type chromosome 17 homolog. It appears likely that t haplotypes evolved directly from a wild-type chromosome through the initial, chance accumulation of linked alleles that together provided a slight distortion of transmission ratio. This small advantage would have been sufficient to serve as the driving force for further structural and functional evolution into the t haplotypes existing today that exhibit very high levels of TRD (Silver 1985; Hammer et al. 1989). Although this evolutionary process resulted in coupled, variant alleles at a number of chromosome 17 loci, the vast majority of genes present within the 20-cM length of t haplotypes have been left unaltered with wild-type functions. There is no reason to expect that t haplotypes provide any selective advantage to the animals in which they reside, since their persistence in wild mice can be accounted for by TRD alone.

The TRD system provides an excellent model for the analysis of normal male germ cell differentiation for two reasons. First, it has been shown that the TRD phenotype is caused not by a simple elimination of gene activity at a single locus, but rather by changes in interactions among multiple gene products. Second, since the mutant phenotype is a quantitative, rather than a qualitative, alteration in fertility, it is easy to develop sensitive assays for subtle effects caused by the incorporation of candidate genes for TRD loci into transgenic animals. The remainder of this paper focuses on the biological, genetic, and molecular strategies that have been employed to understand the phenomenon of TRD, and the identification and functional analysis of a gene that plays a central role in this phenotype.

TRANSMISSION RATIO DISTORTION
Spermatogenesis

Spermatogenesis is the process by which diploid spermatogonial stem cells progressively differentiate into haploid spermatozoa that are capable of fertilizing an egg. Although spermatogenesis is an intricate and complex process, for our purposes, it can be simplified as shown in Figure 2. Each tetraploid spermatocyte undergoes meiosis to give rise to four haploid spermatids. The spermatids undergo complex morphological changes that lead to the characteristic shape and motility of mature sperm. Interestingly, cytokinesis is incomplete throughout differentiation, and spermatids derived from the same stem cell remain linked by cytoplasmic bridges until they are shed into the lumen of the testes tubule. Cytoplasmic bridges allow the passage of at least some mRNA and polypeptide products between developing spermatids (Braun et al. 1989; Caldwell and Handel 1991). By such a process, genetically distinct cells could be rendered functionally equivalent. It has been hypothesized that cytoplasmic bridges function to protect haploid spermatids from the ill effects caused by the potential presence of recessive deleterious alleles.

If cytoplasmic bridges exist as a mechanism to overcome genetic differences between sperm, then TRD is a clear exception to this principle. Heterozygous +/t males produce two classes of sperm, +-bearing and t-bearing, in equal numbers, and yet 95% of their offspring will receive the t chromosome (Silver and Olds-Clarke 1984). Since these sperm are clearly unequal in terms of their contribution to progeny, the implication is that they must be functionally unequal as well. At what level does this functional difference arise?

Physiology

An essential question is whether TRD occurs in +/t males because + sperm are rendered dysfunctional or because t sperm are superior to

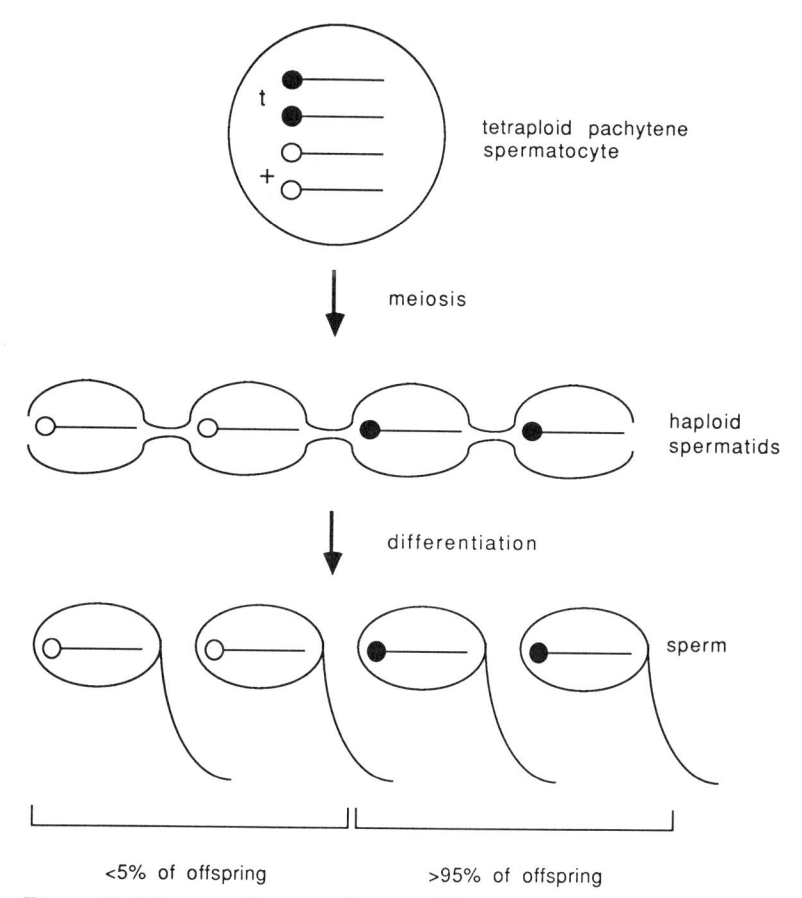

Figure 2 Schematic diagram of sperm differentiation within +/t male mice.

their +-bearing meiotic partners. Two experiments have been performed to address this question. In the first, chimeric male mice were formed by the fusion of $+^A/t$ and $+^B/+^B$ embryos (where $+^A$ and $+^B$ simply represent phenotypically distinguishable wild-type forms), and the relative transmission of the three chromosome 17 types, $+^A$, t, and $+^B$, was analyzed by breeding to wild-type females (Seitz and Bennett 1985). The results demonstrated high levels of transmission of both $+^B$ and t chromosomes, with the near absence of transmission of $+^A$ chromosomes. These results were confirmed and extended by sperm-mixing experiments, in which equal numbers of sperm from +/+ and +/t donors were combined for insemination (Olds-Clarke and Peitz 1985). Again, the + sperm from the +/t donor did not contribute to the progeny, whereas the t sperm and + sperm from the +/+ donor contributed to progeny directly in proportion to their presence in the sperm mixture.

These data provide a first picture of the mechanistic basis for the TRD phenotype. First, the results clearly demonstrate that ejaculated t-bearing sperm have no harmful effect on +-bearing sperm in their vicinity. Second, it is clear that t-bearing sperm are not intrinsically superior to +-bearing sperm. Instead, the presence of a t haplotype in a +/t male must somehow lead to the functional inactivation of +-bearing sperm prior to ejaculation. The results from the chimera experiment provide the best indication of the mechanism by which this inactivation must occur. Spermatids derived from the $+^A/t$ and $+^B/+^B$ genotypes must lie in intimate association within the confines of the chimeric testes, and yet only the $+^A$-containing cells became dysfunctional, whereas $+^B$ spermatids remained unaffected. This result clearly indicates the cytoplasmic bridges as the conduit through which the TRD phenotype must be expressed.

The nature of the + sperm defect may be quite subtle, since it appears that equal numbers of + and t sperm are produced and deposited in the female reproductive tract by +/t males (Silver and Olds-Clarke 1984). One series of studies suggests that the motility of + sperm may be altered in a way that prevents progression to the egg (Olds-Clarke 1983, 1986, 1989). In another study, the acrosome, a sperm organelle that carries lytic enzymes necessary for penetration and fertilization of the egg, was identified as one of the defective components of the dysfunctional + sperm (Brown et al. 1989).

Genetics

A general understanding of the genetic basis for TRD was provided through a series of elegant breeding experiments performed by Mary Lyon (1984, 1986; Lyon and Mason 1977). Lyon took advantage of a set of "partial t haplotypes" that are formed through rare crossovers between "complete t haplotypes" and wild-type forms of the t complex. Partial t haplotypes carry only a portion of the t-DNA present in a complete t haplotype, and most are unable to express the very high levels of TRD associated with the complete haplotypes. Lyon assumed correctly that the loss of the complete TRD phenotype was caused by the loss of t alleles at one or more of the multiple loci that play a role in this phenotype. By combining different partial t haplotypes together within single animals, Lyon was able to "restore" the TRD phenotype. To date, a total of six independent TRD loci have been defined (Fig. 1B) (Lyon 1984; Silver and Remis 1987; Silver 1989).

One of these loci, the t complex responder (Tcr), was found to play a distinctive and central role in the phenotype. The Tcr locus acts in a haploid-specific manner to establish the chromosome 17 homolog that will be transmitted at a high ratio (Fig. 1, genotypes 1 and 2). Distortion from Mendelian ratios was only observed when the *t* allele at this locus was present and heterozygous; males homozygous for either *Tcr+* or *Tcr^t* do not show distortion under any circumstances (Fig. 1, genotypes 3 and 4) (Lyon 1984). The remaining five loci are referred to as *t* complex distorters (*Tcd*s). The *t* alleles at each of the *Tcd* loci can act in *cis* or *trans* configuration relative to *Tcr^t*, and their activity is generally additive and dosage-dependent (Lyon 1986; Silver and Remis 1987; Silver 1989). With the removal of each *Tcd^t* allele, the transmission of *Tcr^t* is reduced. In the extreme case, when a genotype has only a *Tcr^t* allele and no *Tcd^t* alleles, the transmission of *Tcr^t* is reduced to 10–30%, surprisingly below Mendelian expectations (Fig. 1, genotype 5).

Lyon used the same partial *t* haplotypes to study the sterility phenotype characteristic of males with two complete *t* haplotypes (Lyon 1986). For this study, males were bred to homozygosity for certain *t* haplotype regions and to heterozygosity for other haplotype regions. The results of fertility tests led Lyon to hypothesize that the same set of *Tcd^t* alleles responsible for TRD in heterozygotes was also responsible for the sterility phenotype expressed by homozygotes. In contrast, there was no indication that homozygosity for *Tcr^t* by itself reduces fertility.

The accumulated data led Lyon to formulate a working model for TRD (Lyon 1986). First, the sterility experiments suggested that *Tcd^t* alleles are generally deleterious to sperm function. Since these alleles act in *cis* or in *trans* configuration to *Tcr^t*, the implication is that their products must be distributed among all classes of haploid spermatids. This would be accomplished if the *Tcd*s were expressed premeiotically or if they were expressed postmeiotically with unhindered movement of their products through cytoplasmic bridges. Second, the function of the *Tcr^t* allele would be to counteract the deleterious effects of the *Tcd^t* products. The *Tcr^t* allele must be expressed postmeiotically, and its products must be confined to the spermatids that contain that allele. The model accounts for both the sterile phenotype expressed by *t/t* homozygotes and the TRD phenotype expressed by +/*t* heterozygotes. However, it is not completely satisfactory in that it does not explain the reduced transmission characteristic of males that carry an isolated *Tcr^t* allele.

Nevertheless, with the accumulated data, it is possible to set several criteria that must be met by any candidate for *Tcr*. First, the candidate gene must map to the same *t* complex interval in which the *Tcr* locus has been mapped phenotypically. Second, either or both alleles of the candidate gene must express a haploid-specific product in male germ cells. Third, this haploid-specific product(s) must be confined to the postmeiotic cells in which it is expressed. Finally, it is obvious that *Tcr^t* and *Tcr+* must express distinguishable products. With these criteria, it has been possible to narrow the search for the *Tcr* gene to a single candidate, as discussed in detail below.

A *t* COMPLEX RESPONDER LOCUS CANDIDATE

The *Tcp-10b* gene is a candidate for *Tcr*

When a locus is defined by an interesting phenotype but makes an unknown product, the only entry into that locus is through genetics. The basic approach is to use a large set of molecular markers, together with classic breeding studies, to map the locus of interest to a small genomic region which can then be reached molecularly by walking from the nearest cloned markers. The identification of actual candidate genes within this region is then accomplished by scanning the DNA for an appropriate pattern of expression. To pursue the *Tcr* locus in this manner, it was first necessary to generate a panel of molecular markers for use in the mapping studies. At the time this work began, unlike today, no genomic region in the mouse was well marked molecularly. To overcome this problem, a microdissection protocol was used to obtain pure subchromosomal fragments from the *t* complex that could be used as the starting material for generating a panel of region-specific "microclones" (Röhme et al. 1984). These microclones were used in conjunction with partial *t* haplotypes to define physically the *t* complex subregions involved in TRD (Fox et al. 1985). By chance, one clone in particular, Tu66, was found to hybridize to sequences that mapped to multiple sites closely linked to, and surrounding, the *Tcr* locus.

With Tu66 as an initial probe, it has been possible to clone across the region of chromosome 17 that must contain the *Tcr* gene (Schimenti et al. 1987; Rosen et al. 1990). This region is defined to the highest resolution by three partial *t* haplotypes whose recombination breakpoints flank the phenotypically defined *Tcr* locus and lie at a distance of 40–160 kb from each other (Fig. 3). When this region was scanned for testes expression by Northern blot analysis, a single transcription unit was identified. This transcription unit cross-hybridizes with other members of a highly homologous, multigene family—called *t* complex protein-10 (*Tcp-10*)—present in two to four functional copies on different chromosome 17 homologs (Fig. 3) (Schimenti et al. 1988). The *Tcp-10* gene family is expressed exclusively in the testes, and *Tcp-10* RNA is found in spermatogenic cells from the premeiotic pachytene spermatocyte stage onward. The one member of this gene family that maps between the breakpoints described above, *Tcp-10b*, clearly represented an excellent candidate for the *Tcr* locus.

Haploid-specific expression of the *Tcr^t* candidate gene

If the *t* allele of the *Tcp-10b* gene is indeed *Tcr^t*, this member of the *Tcp-10* gene family must be functionally distinct from the other *Tcp-10* genes and alleles that cannot express *Tcr^t* activity. One obvious manner in

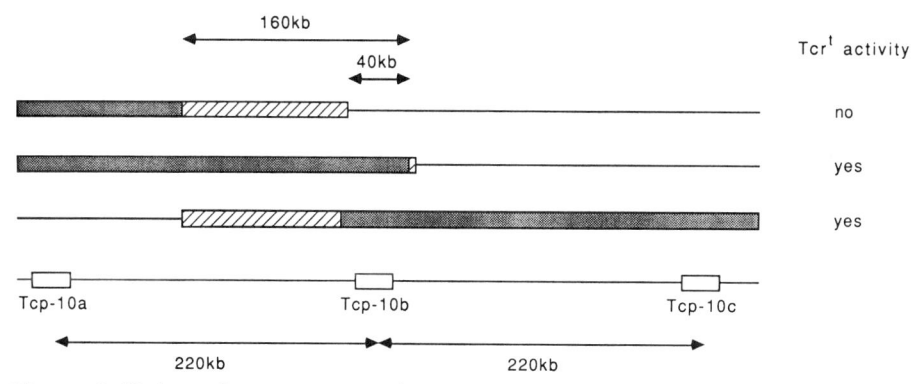

Figure 3 High-resolution mapping of *Tcr*. Three partial *t* haplotypes are shown, with the extent of *t* DNA denoted by shaded boxes and wild-type DNA denoted by lines. The locations of the recombination breakpoints have been resolved to the extent indicated by the hatched boxes. Locations of *Tcp-10* genes are shown below the haplotypes. The presence or absence of *Tcr^t* activity is indicated and was determined in breeding experiments using these haplotypes.

which a functional distinction could arise would be through a difference in primary sequence. However, sequences from the full-length open reading frames of each of the *t* and + members of the gene family show pairwise identities of 98.6% or greater at the nucleotide level (Pilder et al. 1991). Only two consistent amino acid changes distinguish *Tcp-10b^t* from all other *Tcp-10* gene products; one change is nonconservative (Met→Lys), and the other is conservative (Lys→Arg). Although there is abundant evidence that a single amino acid change is sufficient to alter the function of a gene radically, it still seemed possible that a unique feature of *Tcp-10b^t* had not been revealed through our initial analysis of cDNA clones.

As an alternative approach, the polymerase chain reaction (PCR) was utilized to identify all possible transcripts made by each *Tcp-10* gene (Cebra-Thomas et al. 1991). Our strategy was to coamplify the coding region of all *Tcp-10* mRNAs and then identify transcript species made by individual alleles through hybridization to allele-specific oligonucleotides. The results indicated that all of the *Tcp-10* genes, except *Tcp-10b^t*, produce a single transcript. In contrast, *Tcp-10b^t* produces a full-length mRNA and an additional, smaller mRNA transcript not made by any other gene or allele member of the *Tcp-10* family. Analysis of a cDNA clone corresponding to this second "novel" transcript indicates that it is produced through a set of alternative splicing events (Fig. 4). The consequence of the alternative splicing is profound in that the reading frame in the carboxy-terminal portion of the full-length protein is shifted and truncated. Moreover, the altered region is the most conserved portion of the protein in a comparison between humans and mice (S.D. Islam et al., unpubl.). Thus, this second *Tcp-10b^t* transcript fulfills

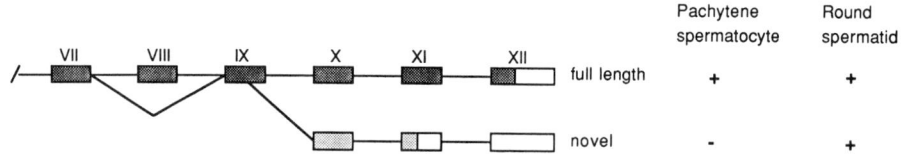

Figure 4 Tcp-$10b^t$ generates two transcripts whose expression differs during spermatogenesis. A portion of the Tcp-$10b^t$ gene structure is drawn schematically to include exons VII–XII. Exons used by the full-length transcript are shaded (top); the novel transcript splices out exon VIII and uses a cryptic splice donor in exon IX to join with the acceptor of exon X (bottom). This last splicing event is out of frame and results in a completely unrelated amino acid sequence, indicated by the lightly shaded boxes at the 3' end of the coding region. Presence or absence of each transcript during stages of spermatogenesis is indicated.

a critical requirement for a Tcr candidate gene by providing an allele-specific product that could clearly be functionally distinct from all other Tcp-10 products.

For the unique Tcp-$10b^t$ transcript to function as expected for a Tcr^t product, it must be expressed in a haploid-specific manner. To determine if this was indeed the case, spermatogenic cells were fractionated on the basis of unit gravity sedimentation in a BSA gradient, and the RNA within each cell fraction was subjected to PCR analysis as described above. The results demonstrate expression of the full-length Tcp-$10b^t$ transcript in both premeiotic cells and haploid cells, whereas the novel, alternatively spliced transcript is expressed only postmeiotically, in the round and elongating spermatids (Fig. 4). All of the other Tcp-10 genes and alleles show an unchanging pattern of transcription throughout pre- and postmeiotic stages. Thus, Tcp-$10b^t$ meets the criteria required of a Tcr^t candidate with respect to map location and pattern of expression. Nevertheless, final proof of identity requires a demonstration of expected function.

Functional analysis of Tcp-$10b^t$

In many cases, it is possible to assess gene function through the use of in vitro tissue culture systems. However, in vitro tests are not sufficient for the analysis of genes defined only through phenotypes expressed within the context of the whole organism. TRD is one such phenotype. Two different approaches have been devised to test gene function within the mouse as a whole organism. The transgenic technology allows one to add genetic material to the genome and is useful in cases where a dominant phenotype is expected. Gene replacement technology is required in those cases where the mutant phenotype is recessive (Capecchi 1989). The TRD phenotype catalyzed by Tcr^t is dominant in that Tcr^t/Tcr^+ heterozygotes express the mutant phenotype. Thus, the transgenic technology is ideal for a determination of the potential Tcr^t activity of the two $Tcp\text{-}10b^t$ transcripts.

Toward this goal, transgenic lines of mice were produced that contained DNA constructs with partially prespliced versions of either the full-length (Tg1) or novel, alternatively spliced (Tg2) transcripts (L.C. Snyder and L.M. Silver, unpubl.). Each construct contained the endogenous $Tcp\text{-}10b^t$ promoter along with genomic DNA comprising the first two exons and introns of the gene. The remaining portion of each construct was derived from corresponding cDNA clones. We chose to study only those lines in which transgene mRNA was expressed in the testes at approximately the correct level. These transgenes were bred into mice with different combinations of partial or complete t haplotypes or wild-type chromosomes 17. Males were progeny tested to determine if the transgene functioned to alter the transmission ratio expected with each genotype. The results obtained in the context of each of four different background genotypes are discussed as follows (Table 1).

Males that are heterozygous for an isolated Tcr^t locus transmit that locus to a reduced percentage (10–30%) of offspring (Fig. 1, genotype 5). Thus, one might expect a functional Tcr^t transgene to be transmitted similarly at a reduced ratio in a +/+ male mouse. However, the results obtained with both the full-length and novel transgenes were the same: No deviation from 50% transmission was observed. A possible explanation for this result comes from the nature of the transgene constructs. As described above, the endogenous $Tcp\text{-}10b^t$ promoter present in each transgene is normally active premeiotically as well as postmeiotically. Thus, one would not expect haploid-specific expression of either of the two prespliced transgenes, and this is the result actually obtained (L.C. Snyder and L.M. Silver, unpubl.). Consequently, the protein product of each transgene may be present in haploid cells that do not contain the transgene as well as in those that do. If all haploid cells have the same Tcr^t products, TRD does not occur (Fig. 1, genotypes 3 and 4).

Similar results were obtained when each transgene was bred into a background that contained two partial t haplotypes with a complete set

Table 1 Effect of transgenes on chromosome 17 TRD

	+/+	+/t*[a]	+/t^{h2}	+/t^{w5}
No transgene[b]	50%	50%	t=30%	t=95%
Tg1	=	=	↑	↓
Tg2	=	=	↓	↑

[a]$t*$ refers to a partial t haplotype that provides all of the Tcd^t loci but has no endogenous Tcr^t locus.
[b]Observed transmission of chromosome 17 from each genotype in the absence of any transgene.
(=) No change in transmission ratio.
(↑) Increased transmission of the t haplotype.
(↓) Decreased transmission of the t haplotype.

of Tcd^t alleles in the absence of a Tcr^t region (Table 1, +/t*). Although an isolated Tcr^t gene bred into such a background would be transmitted at a high ratio (Fig. 1, genotype 2), this was not observed with either of the Tcp-$10b^t$ transgenes. Once again, a likely explanation is that the haploid cells were not biochemically distinct in terms of Tcp-$10b^t$ protein.

Both transgenes, however, showed an effect in the final two t complex backgrounds. These backgrounds differ from the previous two in that both contain an endogenous copy of Tcr^t and both already express a form of TRD in the absence of any transgenes. The partial haplotype t^{h2} contains only a small portion of a complete t haplotype with Tcr^t and a single Tcd^t allele; males heterozygous for t^{h2} transmit it to approximately 30% of offspring (see Table 1). In the presence of the full-length transgene Tg1, the frequency of t^{h2} transmission was increased significantly, to a level closer to 50%; in the presence of the novel transgene Tg2, the transmission of t^{h2} was reduced significantly below 30%. Both transgenes exhibited similar effects in experiments on the $+/t^{w5}$ background. The complete t^{w5} haplotype is normally transmitted to 95% of the offspring. In the presence of the full-length transgene, transmission is reduced toward 50%, whereas in the presence of the novel transgene, t^{w5} transmission is further increased.

The full-length and novel, alternatively spliced transgenes have opposite effects on TRD in transgenic mice. For both t^{w5} and t^{h2}, the presence of the full-length transgene within t-bearing haploid cells leads to a moderation of the TRD effect toward Mendelian transmission levels of 50% (see Table 1). We interpret this to mean that the full-length transcript has wild-type (Tcr^+) activity. In contrast, the presence of the novel, mutant transgene together with an endogenous mutant Tcr^t allele leads to accentuation of TRD away from Mendelian levels in both cases. These results are consistent with the idea that the novel transcript has mutant Tcr^t activity. Thus, we have provided the first functional evidence that Tcp-$10b^t$ is Tcr^t. Moreover, the results also lead to the unexpected conclusion that the Tcp-$10b^t$ allele by itself has dual wild-type and mutant t activities.

MODEL FOR TRANSMISSION RATIO DISTORTION

In light of the transgenic data, how does the Lyon model for TRD hold up? Tcr^t was postulated to be an isolated locus that produced a mutant product with the ability to protect sperm from the harmful effects of the Tcd^t alleles (Lyon 1986). Implicit in the model is the assumption that the Tcr and Tcd products interact in some way, however indirect that may be. This model still presents a valid framework for thinking about TRD, but the real picture is now more complicated because Tcr is just a single member of a multigene family, and it appears that the mutant Tcr^t allele by itself can produce both wild-type and mutant products. When this new finding is considered together with the fact that Tcr^t can cause either an increase or a reduction in transmission ratio, depending on the presence of Tcd^t alleles, it seems reasonable to suggest that the interaction between the products of the various genes is probably competitive in that concentrations and affinities of the various gene products determine the direction and strength of the final TRD phenotype. If the abundance of one or more of these gene products is altered, as in the transgenic experiments described here, the final level of TRD is altered accordingly.

The TRD model of Bullard and Schimenti (1990) provides the closest fit to the observed results. This model still assumes that Tcd products can flow freely between haploid cells, whereas Tcr^t products remain within the haploid cells that produce them. The new features of this model are that (1) the Tcr^+ products will be present in all haploid cells and (2) a wild-type $Tcr^+ \leftrightarrow Tcd^+$ complex is essential for proper sperm function. Other $Tcr \leftrightarrow Tcd$ combinations can be formed, but they are all nonfunctional. We would like to add to this model by suggesting that the one functional combination has the least relative affinity (Fig. 5A).

This revised TRD model accounts for all of the results obtained in breeding studies with mice that carry different combinations of complete and partial t haplotypes. For example, the role played by the mutant Tcr^t product in sperm bearing complete t haplotypes would be to sequester mutant Tcd^t products and thus facilitate the formation of functional $Tcr^+ \leftrightarrow Tcd^+$ complexes (see Fig. 1, genotype 1; Fig. 5B). The wild-type meiotic partners of t-bearing sperm would not have this protection, and the mutant Tcd^t products would compete more effectively for binding to the Tcr^+ product in these cells, resulting in dysfunctional + sperm.

In contrast, Tcr^t would have an opposite effect in mice that carried this allele in the absence of Tcd^t alleles (Fig. 1, genotype 5; Fig. 5C). In this case, Tcr^t products would sequester Tcd^+ products and thus prevent the formation of functional $Tcr^+ \leftrightarrow Tcd^+$ complexes in the sperm that carried the Tcr^t allele. The +-bearing sperm from these animals would be unaffected and would retain full function.

In the context of this model, it is possible to explain both the moderating and accentuating effects of the two transgenes on TRD through shifts in the equilibrium concentrations of each $Tcr \leftrightarrow Tcd$ complex. In the $+/t^{h2}$ experiment (see Fig. 5C), the full-length transgene, with Tcr^+ activity, may moderate t^{h2} transmission because increased Tcr^+ concentrations would favor the functional $Tcr^+ \leftrightarrow Tcd^+$ interaction in t^{h2} sperm. In contrast, the novel transgene reduces t^{h2} transmission further. Although this construct contributes Tcr^t activity to all sperm, enough Tcr^t activity may be present in the t sperm to further diminish the

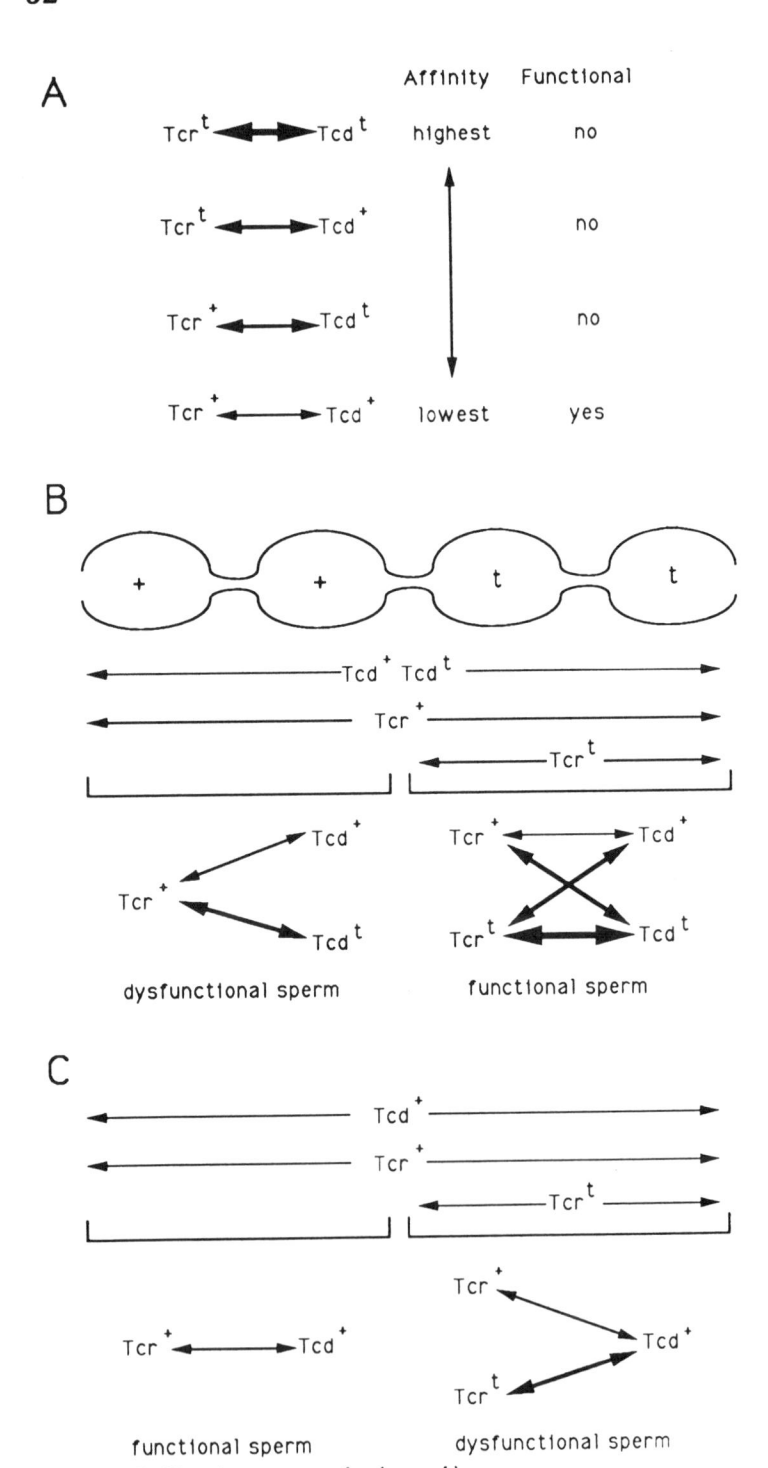

Figure 5 (See facing page for legend.)

favorable $Tcr^+ \leftrightarrow Tcd^+$ interaction, resulting in relatively less functional t sperm.

The behavior of the transgenes in the presence of the complete t haplotype t^{w5} can also be addressed (see Fig. 5B). The full-length construct may contribute to the favorable $Tcr^+ \leftrightarrow Tcd^+$ interaction in the + sperm, which may improve their function, and/or the construct may contribute to unfavorable $Tcr^+ \leftrightarrow Tcd^t$ interactions in the t sperm, making them less functional. The novel construct contributes more Tcr^t which, although present in all sperm, again may reach a critical level in t sperm to inactivate all Tcd^t factors and make the t sperm relatively more functional than the + sperm.

FINAL THOUGHTS AND FUTURE DIRECTIONS

Much remains to be learned about the involvement of $Tcp\text{-}10b^t$ in TRD. One interesting aspect of this work is the alternative splicing that $Tcp\text{-}10b^t$ demonstrates only in haploid spermatids, although the gene is transcribed earlier in pachytene spermatocytes. This result implies that the testes, and postmeiotic cells in particular, may possess a splicing machinery with altered or relaxed activity such that otherwise cryptic splice donors and acceptors are recognized. In accord with this idea, a minor change in an intronic sequence of $Tcp\text{-}10b^t$ has been identified, which could be responsible for the generation of the novel transcript from this otherwise normal member of the $Tcp\text{-}10$ family (K. Lukitsch, unpubl.).

The actual function of the TCP10 proteins remains one of the most crucial problems yet to be resolved, as it could provide important insight into the developmental processes affected during spermatogenesis in +/t males. Unfortunately, the amino acid sequence does not reveal much about its possible behavior, because there is no significant homology

Figure 5 A model for TRD. (A) Predicted affinity and function of $Tcr \leftrightarrow Tcd$ complexes. For simplicity, Tcd^t and Tcd^+ represent all t or wild-type Tcd gene products, respectively. Heavier arrows indicate greater affinity or strength of interaction. (B,C) Predicted $Tcr \leftrightarrow Tcd$ interactions in spermatids produced by a heterozygous +/t male mouse. Spermatids bearing +- or t-chromosome 17 are linked by cytoplasmic bridges. Products of the Tcd and Tcr^+ loci are assumed to be present in all cells, whereas Tcr^t products are restricted to t-sperm. In B, we show the interactions occurring in a male mouse heterozygous for a complete t haplotype, as in Fig. 1, genotype 1. In +-bearing sperm, the stronger, nonfunctional $Tcr^+ \leftrightarrow Tcd^t$ interaction is favored, resulting in dysfunctional sperm. In t-bearing sperm, the most favored interaction is $Tcr^t \leftrightarrow Tcd^t$, leaving Tcr^+ and Tcd^+ free to interact and resulting in functional sperm. In C, we show the interactions occurring in spermatids produced by a male mouse carrying a partial t haplotype with an isolated Tcr^t locus, as in Fig. 1, genotype 5. In t-bearing sperm, the stronger, nonfunctional $Tcr^+ \leftrightarrow Tcd^t$ interaction is favored, resulting in dysfunctional sperm. The + sperm only contain the favorable $Tcr^+ \leftrightarrow Tcd^+$ interaction, which results in functional sperm.

with other sequences in the Genbank database. Nor does the protein contain sequence motifs indicative of particular functions or structures, such as signal peptides, helix-loop-helix domains, and transmembrane domains. The one clue at this time is that the carboxy-terminal 180 amino acids comprise 20 nearly perfect repeats of a 9-amino-acid motif. The repeat motif is highly conserved between human and mouse (S.D. Islam et al., unpubl.), suggesting that it may contribute to a wild-type function conserved between these species. This motif is the region of the protein removed by the *Tcp-10b^t* alternative splice event, suggesting that the polypeptide made by the novel transcript could have *Tcr^t* activity because it has a subset of *Tcr$^+$* activities or because it has acquired a new function. Initial immunoprecipitation and immunofluorescence studies using antibodies directed against a synthetic TCP10 protein suggest that the full-length protein may be associated with internal membranes (J.A. Cebra-Thomas et al., unpubl.), some of which could be integral to the formation of the acrosome (Guraya 1987). These data may be significant in view of the biological studies suggesting that the acrosome may be one structure affected in +/*t* mice. A second crucial area that must be investigated is one on which models for TRD are based—the assumption that the *Tcr^t* product acts in a haploid fashion and is present only in sperm which contain that gene. Antibodies specific for the novel TCP10bt polypeptide should enable us to address this question.

Other critical pieces of the TRD puzzle are the identities of the *Tcd* loci and the nature of their interactions with *Tcr* in +/+ and +/*t* male mice. Currently, there are three candidates for *Tcd* loci based on map locations and pattern of expression in the testes: *Tctex-1* for *Tcd-1*, *Tcte-3* for *Tcd-3*, and *Tcp-1* for *Tcd-4*. If any of these candidates are correct, one would again expect differences in the primary sequence of their *t* and + alleles, and a functional association of the *t* form with sterility in homozygous animals. One difficulty in identifying *Tcd-2* candidates is that this locus has been assigned to a very large chromosomal region (Silver 1989). The effect of *Tcd-2* in TRD could therefore result from the action of one locus or many, illustrating a common difficulty in moving from genetically defined loci to molecular candidates.

The ultimate goal of many who study the *t* complex is a fuller understanding of the processes of spermatogenesis and development. Given that 324 loci have been mapped to chromosome 17 (Committee for Mouse Chromosome 17 1991), candidates for the lethal *t* complex loci, as well as those involved in TRD, are certain to be identified and characterized in the near future. It can be inferred that knowledge obtained through the analysis of these loci will be applicable to human biology as well. Although we know that the *t* complex does not exist as an intact unit in humans (Willison et al. 1987; Bibbins et al. 1989), wild-type homologs of the genes causing TRD are almost certainly present (S.D. Islam et al., unpubl.). Since some of the loci involved in TRD are associated with reduced fertility in mice, it may be that mutations in these genes are associated with human male infertility, in at least some cases. Further study of this intriguing region of the mouse genome is sure to be fruitful in the coming years as we continue to explore the relationship between phenotype and genotype.

Acknowledgments

We thank our colleagues in the Silver and Tilghman laboratories for helpful discussions and reviews of the manuscript. Research reported from our laboratory was supported by grants from the National Institutes of Health to L.M.S. and a postdoctoral fellowship from the National Institutes of Health (HD-07173) to L.C.S.

References

Artzt, K., P. McCormick, and D. Bennett. 1982a. Gene mapping within the T/t complex of the mouse. I. *t*-lethal genes are nonallelic. *Cell* **28**: 463.

Artzt, K., H.-S. Shin, and D. Bennett. 1982b. Gene mapping within the T/t-complex of the mouse. II. Anomalous position of the *H-2* complex in *t*-haplotypes. *Cell* **28**: 471.

Bibbins, K.B., J.-Y. Tsai, J. Schimenti, N. Sarvetnick, H.Y. Zoghbi, P. Goodfellow, and L.M. Silver. 1989. Human homologs of two testes-expressed loci on mouse chromosome 17 map to opposite arms of chromosome 6. *Genomics* **5**: 139.

Braun, R.E., R.R. Behringer, J.J. Peschon, R.L. Brinster, and R.D. Palmiter. 1989. Genetically haploid spermatids are phenotypically diploid. *Nature* **337**: 373.

Brown, J., J.A. Cebra-Thomas, J.D. Bleil, P.M. Wassarman, and L.M. Silver. 1989. A premature acrosome reaction is programmed by mouse *t* haplotypes during sperm differentiation and could play a role in transmission ratio distortion. *Development* **106**: 769.

Bullard, D.C. and J.C. Schimenti. 1990. Molecular cloning and genetic mapping of the *t* complex responder candidate gene family. *Genetics* **124**: 957.

Caldwell, K.M. and M.A. Handel. 1991. Protamine transcript sharing among postmeiotic spermatids. *Proc. Natl. Acad. Sci.* **88**: 2407.

Capecchi, M. 1989. The new mouse genetics: Altering the genome by gene targeting. *Trends Genet.* **5**: 70.

Cebra-Thomas, J.A., C. Decker, L.C. Snyder, S.H. Pilder, and L.M. Silver. 1991. Allele- and haploid-specific product generated by alternative splicing from a mouse *t* complex responder locus candidate. *Nature* **349**: 239.

Committee for Mouse Chromosome 17. 1991. Mouse chromosome 17: Second report. *Mammalian Genome* **1**: 280.

Dobrovolskaia-Zavadskaia, N. 1927. Sur la mortification spontanée de la queue chez la souris nouveau-née et sur l'existence d'un caractère (facteur) héréditaire. *C.R. Seances Soc. Biol.* **97**: 114.

Dobrovolskaia-Zavadskaia, N. and N. Kobozieff. 1932. Les souris anoures et a queue filiforme qui se reproduisent entre elles sans disjonction. *C.R. Seances Soc. Biol.* 110: 782.

Fox, H.S., G.R. Martin, M.F. Lyon, B. Herrmann, A.-M. Frischauf, H. Lehrach, and L.M. Silver. 1985. Molecular probes define different regions of the mouse *t* complex. *Cell* 40: 63.

Guraya, S.S. 1987. *Biology of spermatogenesis and spermatozoa in mammals.* Springer-Verlag, Berlin.

Hammer, M.F., J. Schimenti, and L.M. Silver. 1989. Evolution of mouse chromosome 17 and the origin of inversions associated with *t* haplotypes. *Proc. Natl. Acad. Sci.* 86: 3261.

Herrmann, B., M. Búcan, P.E. Mains, A.-M. Frischauf, L.M. Silver, and H. Lehrach. 1986. Genetic analysis of the proximal portion of the mouse *t* complex: Evidence for a second inversion within *t* haplotypes. *Cell* 44: 469.

Hogan, B., F. Costantini, and E. Lacy, eds. 1986. *Manipulating the mouse embryo: A laboratory manual.* Cold Spring Harbor Laboratory, Cold Spring Harbor, New York.

Klein, J., P. Sipos, and F. Figueroa. 1984. Polymorphism of *t*-complex genes in European wild mice. *Genet. Res.* 44: 39.

Lyon, M.F. 1984. Transmission ratio distortion in mouse *t*-haplotypes is due to multiple distorter genes acting on a responder locus. *Cell* 37: 621.

———. 1986. Male sterility of the mouse *t*-complex is due to homozygosity of the distorter genes. *Cell* 44: 357.

Lyon, M.F., and I. Mason. 1977. Information on the nature of *t*-haplotypes from the interaction of mutant haplotypes in male fertility and segregation ratio. *Genet. Res.* 29: 255.

Olds-Clarke, P. 1983. Nonprogressive sperm motility is characteristic of most complete *t* haplotypes in the mouse. *Genet. Res.* 42: 151.

———. 1986. Motility characteristics of sperm from the uterus and oviducts of female mice after mating to congenic males differing in sperm transport and fertility. *Biol. Reprod.* 34: 453.

———. 1989. Sperm from $t^{w32}/+$ mice: Capacitation is normal, but hyperactivation is premature and non-hyperactivated sperm are slow. *Dev. Biol.* 131: 475.

Olds-Clarke, P., and B. Peitz. 1985. Fertility of sperm from *t*/+ mice: Evidence that +-bearing sperm are dysfunctional. *Genet. Res.* 47: 49.

Pilder, S.H., C.L. Decker, S. Islam, C. Buck, J.A. Cebra-Thomas, and L.M. Silver. 1991. Concerted evolution of the mouse *Tcp-10* gene family: Implications for the functional basis of *t* haplotype transmission ratio distortion. *Genomics* (in press).

Röhme, D., H. Fox, B. Herrmann, A.-M. Frischauf, J.-E. Edström, P. Mains, L.M. Silver, and H. Lehrach. 1984. Molecular clones of the mouse *t* complex derived from microdissected metaphase chromosomes. *Cell* 36: 783.

Rosen, L.L., D.C. Bullard, L.M. Silver, and J.C. Schimenti. 1990. Molecular cloning of the *t* complex responder genetic locus. *Genomics* 8: 134.

Sarvetnick, N., H. Fox, E. Mann, P. Mains, R. Elliot, and L.M. Silver. 1986. Nonhomologous pairing in mice heterozygous for a *t* haplotype can produce recombinant chromosomes with duplications and deletions. *Genetics* 113: 723.

Schimenti, J., L. Vold, D. Socolow, and L.M. Silver. 1987. An unstable family of large DNA elements in the center of the mouse *t* complex. *J. Mol. Biol.* **194:** 583-594.

Schimenti, J., J.A. Cebra-Thomas, C. Decker, S. Islam, S.H. Pilder, and L.M. Silver. 1988. A candidate gene family for the mouse *t complex responder* (*Tcr*) locus responsible for haploid effects on sperm function. *Cell* **55:** 71.

Seitz, A.W., and D. Bennett. 1985. Transmission distortion of *t*-haplotypes is due to interactions between meiotic partners. *Nature* **313:** 143.

Silver, L.M. 1985. Mouse *t* haplotypes. *Annu. Rev. Genet.* **19:** 179.

———. 1989. Gene dosage effects on transmission ratio distortion and fertility in mice that carry *t* haplotypes. *Genet Res* **54:** 221.

Silver, L. and P. Olds-Clarke. 1984. Transmission ratio distortion of mouse *t* haplotypes is not a consequence of wild type sperm degeneration. *Dev. Biol.* **105:** 250.

Silver, L.M. and D. Remis. 1987. Five of the nine genetically defined regions of mouse *t* haplotypes are involved in transmission ratio distortion. *Genet. Res.* **49:** 51.

Willison, K., A. Kelly, K. Dudley, P. Goodfellow, N. Spurr, V. Groves, P. Gorman, D. Sheer, and J. Trowsdale. 1987. The human homologue of the mouse *t*-complex gene, TCP1, is located on chromosome 6 but is not near the HLA region. *EMBO J.* **6:** 1967.

Cloning the Mammalian Sex-determining gene, *TDF*

Peter N. Goodfellow, J. Ross Hawkins, and Andrew H. Sinclair

Laboratory of Human Molecular Genetics
Imperial Cancer Research Fund
London, WC2A 3PX, United Kingdom

In this chapter, we consider the problems of cloning and identifying genes of unknown biochemical function. The general approach exploits the chromosomal location of the target gene and has been called variously "reverse genetics" or "positional cloning." These now-classic methods are illustrated by reference to the cloning of the mammalian sex-determining gene, *TDF*.

The main topics discussed include:

❏ sex determination in mammals, the central role of the testis, and the genetic contribution of the Y chromosome; structure-function relationships for the Y chromosome and the different roles of the pseudoautosomal and Y-specific regions

❏ the construction and limitations of maps of the human Y chromosome pinpointing the position of *TDF*: meiotic maps of the pseudoautosomal region; deletion or fragment maps of the Y-specific region; and long-range restriction maps linking the meiotic and deletion maps

❏ chromosome walking and searching cloned sequences for genes

❏ testing for equivalence between candidate and target genes

INTRODUCTION

There are many different ways to clone a gene; however, the options become more limited if the gene of interest is mammalian and its protein product is unknown. A general solution to the problem of cloning mammalian genes of unknown function is based on a knowledge of the gene's chromosomal position. Originally, this approach was called "reverse genetics," but fashions change, and the more descriptive name of "positional cloning" is currently preferred. In principle, "positional cloning" is straightforward and can be broken down into a series of simple steps:

Step 1. Construction of a map of the relevant region of the genome containing at least three markers: the target gene and two flanking markers
Step 2. Chromosome walking to clone all of the DNA that spans the interval between the flanking markers
Step 3. Searching for genes in the cloned DNA
Step 4. Testing, by correlation, the candidate genes and the target gene
Step 5. Proof of identity between candidate gene and target gene

As with all science, the description of the ideal and the practice can be very different, and it is the differences that we emphasize in this paper. Much of the work we describe has been performed in collaboration with R. Lovell-Badge and colleagues from the National Institute of Medical Research.

THE BACKGROUND OF SEX DETERMINATION

As befits a topic that sits at the interface between genetics and developmental biology, each field has contributed a rule of mammalian sex determination. The first rule asserts the central role of the testis and derives from a series of remarkable experiments performed by A. Jost (Jost et al. 1973). If a male rabbit is castrated in utero early in development, subsequent differentiation is female; if a female rabbit is castrated at the same developmental stage, subsequent differentiation is also female. From these results, it was deduced that the basic developmental plan is female and that the formation of testes changes the fate of the fetus. Further experiments showed that the testes produce hormones and other diffusible products (e.g., testosterone and anti-Mullerian hormone) that are responsible for inducing male secondary sexual characteristics. These experiments allowed the problem of sex determination in mammals to be reduced to the choice between testis and ovary formation.

The second rule of mammalian sex determination asserts that the Y chromosome is a dominant genetic regulator of testis formation. With the advent of better techniques to analyze human chromosomes, it became possible to study chromosome aneuploidies (individuals with unusual numbers of chromosomes). The usual sex chromosome complement in humans and other mammals is two X chromosomes in females and one X chromosome and one Y chromosome in males. Ford et al. (1959) found that individuals with Turner's syndrome have a single X chromosome and are female, implying that the Y chromosome carries a gene (or genes) that is required for testis formation. This conclusion was reinforced by analysis of individuals with Klinefelter syndrome; these males have two X chromosomes and a Y chromosome (Jacobs and Strong 1959). The uncertain nature of the genetic material on the Y chromosome required for male sex determination persuaded human geneticists to name these elements the *t*estis *d*etermining *f*actor (*TDF*); mouse geneticists named the murine homolog *t*estis *d*etermining *Y* gene (*Tdy*). There has been confusion in some quarters, apparently induced by the name, that all of the genetic information required for testis formation is subsumed within *TDF/Tdy*. In our opinion, this is simplistic: Sex determination is under the control of a genetic pathway, and most of the genes are likely to be autosomal or X-linked. In the absence of *TDF/Tdy*, the pathway does not function to induce testis formation; in the presence of *TDF/Tdy*, the pathway is complete for male sex determination. This role has been variously described as a "switch" or a "rate-limiting step."

Half a century ago, Koller and Darlington (1934) peered down a microscope at rat chromosomes undergoing male meiosis. They observed that the X and Y chromosomes were dissimilar in morphology but were able to pair along part of their lengths. They proposed that the Y chromosome was composed of two regions with distinct functions: a region shared between the X and Y chromosomes and a Y-specific region. The shared region would be required in male meiosis for correct pairing and segregation of the sex chromsomes. Maintenance of the shared region would necessitate recombination between the sex chromosomes. The inheritance pattern of markers in the shared region would simulate autosomal behavior (subsequently the term pseudoautosomal was introduced to describe this unique genetics [Burgoyne 1982] and the shared region is also known as the pseudoautosomal region). The Y-linked, sex-specific region would encode genes needed for male sex determination, and recombination must not occur in this region. These clear-sighted predictions were confirmed when the tools of molecular biology were brought to bear on the genetics of the human Y chromosome. For more information on the genetics of sex chromosomes and the biology of sex determination, see Goodfellow et al. (1985) and Goodfellow and Darling (1988).

STEP 1: CONSTRUCTION OF MAPS

A meiotic map of the pseudoautosomal region

The first molecular support for the existence of the pseudoautosomal region (PAR) was the mapping of the gene *MIC2* to the human X and Y chromosomes (Goodfellow et al. 1983). *MIC2* encodes a ubiquitously expressed cell-surface antigen that may have a role in cell-cell recognition. Proof of the existence of the PAR came from the cloning of DNA sequences derived from the Y chromosome. Several sequences were found to map to both the X and Y chromosomes and to recombine between the sex chromosomes (Cooke et al. 1985; Goodfellow et al. 1986; Rouyer et al. 1986). This allowed the construction of a recombination map based on male meiosis (Fig. 1). The map has a number of unique features (Weissenbach et al. 1987):

1. Approximately 50% recombination occurs between markers at the telomere and the pseudoautosomal boundary.
2. Double recombinants have not been observed.
   (Features 1 and 2 imply that an obligate recombination occurs in the PAR in every male meiotic event.)
3. In female meiosis, only a few percent recombination is found in the PAR.
4. The physical map of the PAR is 2.6 Mb in length (Brown 1988; Petit et al. 1988).
   (Features 1 and 4 mean that the rate of recombination in male meiosis is about 20 times the genome average. Features 3 and 4 imply that recombination in female meiosis [or in the female] is close to the genome average.)
5. The physical map, deduced from long-range restriction mapping, and the meiotic map are colinear (Brown 1988; Petit et al. 1988).

MIC2 is located immediately adjacent to the sex-specific region and only recombines with sexual phenotype in about 2% of meioses (Goodfellow et al. 1986; Weissenbach et al. 1987).

A deletion map of the sex-specific region

Sex-reversed individuals apparently break the rule that the Y chromosome "controls" sex determination. The most common class of sex-reversed individuals are XX males, i.e., phenotypic males with testes who lack a Y chromosome. In 1983, Bishop, Weissenbach, and colleagues found Y-derived sequences in the genomes of some XX males

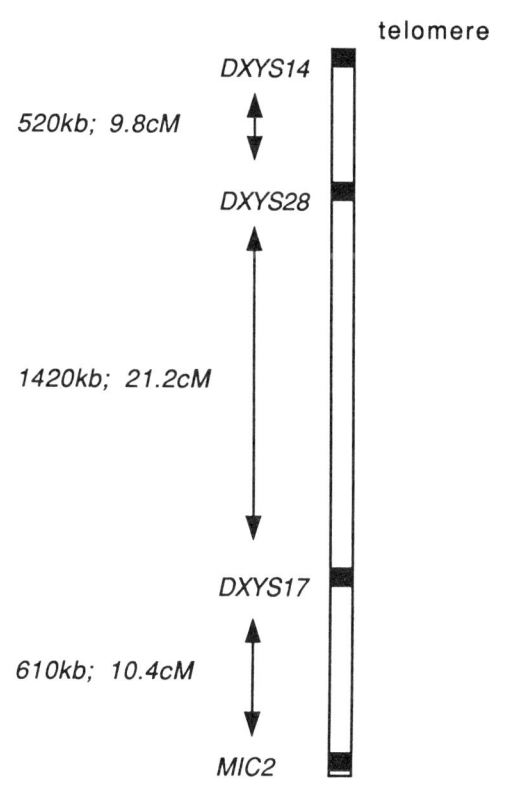

```
                    telomere
      DXYS14      ▪
                  ┃
520kb; 9.8cM   ↕  ┃
                  ┃
      DXYS28      ▪
                ▲ ┃
                  ┃
                  ┃
1420kb; 21.2cM    ┃
                  ┃
                  ┃
                ▼ ┃
      DXYS17      ▪
                ▲ ┃
610kb; 10.4cM   ↕ ┃
                ▼ ┃
      MIC2        ▪
            pseudoautosomal boundary
```

Figure 1 Genetic and physical maps of the pseudoautosomal region. The map has been adapted from the family data summarized in Weissenbach et al. (1987) and the physical data in Brown (1988) and Petit et al. (1988). We have made several assumptions in generating this map; any errors are due to our assumptions and not due to the original authors.

(Guellaen et al. 1984). These sequences had been inherited with the paternal X chromosome and were X-located because of an aberrant recombination between the X and Y chromosomes during male meiosis (Ferguson-Smith 1966). Different XX males inherit different terminal fragments of the Y chromosome, and this allowed the construction of "deletion" or "fragment" maps of the Y chromosome (Vergnaud et al. 1986). The terminal position of *TDF* is inherent in the construction of these maps, and any marker located on the telomere, or distal, side of *TDF* must be present in the genomes of all XX males caused by a terminal exchange between the sex chromosomes. Two conclusions can be drawn from the deletion maps of the Y-specific region: First, *TDF* behaves as a single genetic entity and second, *TDF* maps to a distal position on the short arm of the Y chromosome adjacent to the PAR.

Because of the way the fragments are generated by terminal exchange, it was not possible to estimate accurately the distance between *MIC2*, the distal flanking marker, and *DXYS5Y*, a marker defined by a DNA probe, on the centromeric or proximal side. The best estimates we could make suggested a distance of between 0.5 Mb and 2.0 Mb. There were also a small number of markers that were known to be located close to *TDF* but could not be positioned on one side or the other.

Aberrant terminal exchange between the sex chromosomes can also generate Y chromosomes deleted for *TDF*; individuals that inherit such chromosomes will be sex-reversed XY females. These terminal deletions are usually large and have not been useful for defining the location of *TDF* (Levilliers et al. 1989). However, a smaller deletion was found in an XY female, WHT1013, associated with a Y-22 translocation (Page et al. 1987, 1990).

A long-range restriction map

The meiotic and deletion maps of the Y chromosome are constructed on different principles and cannot be directly joined. In an attempt to circumvent this problem, a long-range restriction map rooted in *MIC2* was constructed (Pritchard et al. 1987). This map spanned the PAR and Y-specific region and identified two landmarks of interest. By comparing the restriction map of the X and Y chromosomes, it was possible to define the position of the boundary of the PAR. The second landmark was an HTF island (CpG-rich region) located in the Y-specific region about 100 kb from the boundary. Because HTF islands are frequently associated with genes, the reasonable prediction was made that this island defined a new gene. Less reasonable was the flight of fancy that suggested the HTF island might mark *TDF* (Pritchard et al. 1987).

Step 1 summarized

TDF was localized to the distal end of the Y chromosome short arm in the Y-specific region. *MIC2* was the distal flanking marker for *TDF*, and *DXYS5Y* was the closest published proximal marker. Other unpublished markers were available in various laboratories. It should be noted that the long-range restriction map did not extend all the way from *MIC2* to *DXYS5Y*. This meant that the distance between the flanking markers was unknown, and a chromosomal walk required stepping into the dark.

STEP 2: CHROMOSOME WALKING

In theory, chromosome walking is simple. Starting with a DNA probe, genomic libraries are screened, and hybridization-positive clones are isolated. The new clones are restriction-mapped, and end fragments are identified to produce new probes to continue the walk. Initially, it is necessary to continue in both directions until a landmark allows orientation. Unfortunately, biology cannot be constrained by simple ideas, and chromosome walking is beset with obstacles to progress. Notable complications include repetitive sequences and unclonable sequences.

Two groups have published chromosome walks across parts of the sex-determining region of the Y chromosome. The first walk was carried out by Page et al. (1987). These authors started their walk from the region proximal to *TDF* and continued to the marker *DYS13*(GMGY3). The latter marker was positioned on the long-range restriction map 100 kb from the boundary. This walk was later extended to the pseudoautosomal boundary (Page et al. 1987, 1990).

The second walk was initiated in *MIC2*, crossed the pseudoautosomal boundary, continued past the markers *DYS104* and *DYS13*, and finished 5 kb short of the sex-specific HTF island on the Y chromosome previously identified as marking a candidate gene for *TDF* (Ellis et al. 1989). This walk was 250 kb long, required about 30 steps, and took over 3 person-years of effort. The reasons for our inefficiency included both inexperience and bad luck:

1. The first libraries we screened were amplified phage and cosmid genomic libraries. Amplified libraries are less representative than primary libraries. To do a chromosome walk, we recommend making one's own very large primary genomic libraries.
2. The walk was through a region that contained a very high level of repetitive sequences. This made identification of suitable probes for continuing the walk very difficult. Several techiques were used to ameliorate this problem, including suppression of repeats by hybridization with total human genomic DNA (this helped with high-copy repeats) and construction of libraries from human-rodent somatic cell hybrids containing the Y chromosome as the only human contribution, thereby reducing the number of repeats in the target library (this was especially useful when confronted with low-copy repeats). For much of the time, both techniques were used together.
3. There have been several duplications and rearrangements of the human Y chromosome during evolution. Even when we managed to isolate what appeared to be "unique" sequences, they frequently hybridized to a second locus on the long arm of the Y chromosome. This caused confusion as different clones "jumped" between the long arm and short arm of the Y chromosome.
4. Yeast artificial chromosomes (YACs) had not been invented (Burke et al. 1987).

Step 2 summarized

During the walk generated by Page et al. (1987), four landmarks were passed (see Fig. 2): (1) the Y chromosome breakpoint in XX male LGL203, which oriented the map and provided a proximal limit to the position of *TDF*; (2) the HTF island present on the long-range restriction map in the Y-specific region; (3) the distal breakpoint in the XY female with a small deletion, which provided the distal limit to the position of *TDF* (WHT1013); and (4) *DXYS13*, another marker that had been positioned on the long-range restriction map, where the walk was stopped.

These results localized *TDF* to a 140-kb region. Within this region, and positioned adjacent to the HTF island, is located the gene *ZFY*.

ZFY, a gene of unknown function

ZFY initially appeared to have several features that might be expected for *TDF* (Page et al. 1987); it is conserved and found on the Y chromosome of all eutherian mammals tested and encodes a protein containing 13 "zinc finger" motifs, a nuclear localization signal, and an activation domain—all features found in transcription factors. Less consistent with a role in sex determination were the findings that *ZFY* is expressed in all human tissues in fetal and adult males; a closely related

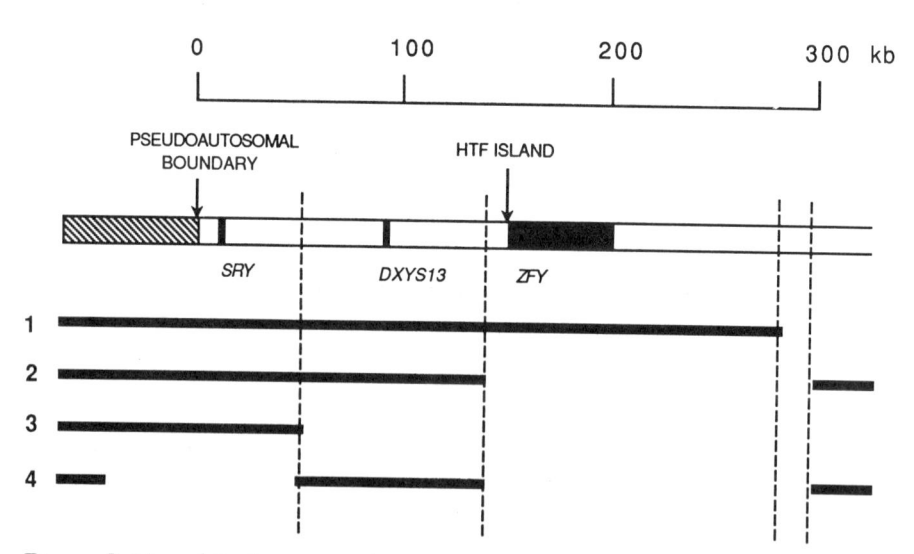

Figure 2 Map of the human sex-determining region. The hatched region represents the pseudoautosomal region; the open bar is Y-specific. The fragments listed underneath are the portions of the Y chromosome: (*1*) XX male LGL203 (Page et al. 1987); (*2*) original interpretation of the XY female WHT1013 (X,t[Y;22]) (Page et al. 1987); (*3*) XX males, ZM, BM, DC, and TL (Palmer et al. 1989); (*4*) XY female WHT103 (Page et al. 1990). Both the Page and Goodfellow groups cloned this entire region in a Y chromosome walk.

gene, *ZFX*, is present on the X chromosome of all eutherian mammals (Schneider-Gadicke et al. 1988); and homologs of *ZFY* are not present on the sex chromosomes of metatherian mammals (marsupials) (Sinclair et al. 1988). The results of the latter experiment prompted us to reevaluate the mapping data that pinpointed the location of *TDF*.

A deletion map of the Y-specific region: Reprise

It was difficult to imagine that the proximal breakpoint positioning *TDF* was erroneous; however, alternative explanations for the sex reversal found in patient WHT1013, with the Y:22 translocation, could be envisaged. For example, the sex reversal might have been due to a position effect on *TDF* caused by the deletion. In this case, *TDF* might map immediately adjacent to the pseudoautosomal boundary.

We decided to test the hypothesis that *ZFY* was not *TDF* by investigating the genomes of *ZFY*-negative XX males. It had previously been reasoned that these individuals were male because of "gain of function mutations" in "downstream" genes in the sex determination pathway; such mutations have been found in many other genetic pathways. Fellous (Palmer et al. 1989) had made a collection of 14 *ZFY*-negative XX males and intersex individuals; among these cases, we found 4 who had inherited Y sequences derived from the region adjacent to the boundary. Although 2 of the patients had an intersex phenotype, the other 2 were male; all 4 had testicular material. These results formally excluded *ZFY* as a candidate for *TDF* and mapped the sex-determining gene to a region of only 60 kb; subsequently, the breakpoints were more precisely mapped to 35 kb from the boundary (Sinclair et al. 1990). The discrepancy between this newly defined sex-determining region and the previously defined region was resolved when a second deletion extending 45 kb into the Y-specific region was found in the XY female WHT1013 (Page et al. 1990). The map of the sex-determining region is summarized in Figure 2.

STEP 3: SEARCHING FOR GENES

Finding genes in long stretches of cloned DNA is probably the limiting step in positional cloning. There is no robust approach that will guarantee success. Those who have followed the logic of this review so far will have realized that *TDF* is located within sequences that we had previously cloned during the chromosome walk from *MIC2*. Initially we had searched this region by the following methods:

> 1. Cross-hybridization to the genomes of other species using the probes isolated during the chromosome walk. This approach

assumes that protein coding sequences are more conserved than noncoding sequences.
2. Screening cDNA libraries, derived from testes and murine genital ridges, with whole cosmids and phage clones.

Neither approach was successful: The cross-hybridization was confounded by choice of probe and the presence of repeats that are shared by human and mouse; the library screening failed because the libraries screened did not contain the relevant cDNA clones. It is also likely that our judgment was colored by the expectation that *TDF* was located at the HTF island identified by the long-range restriction mapping.

However, on the second search the target 35 kb of DNA was subcloned into fragments of about 4 kb and digested with frequent cutter enzymes such as *Rsa*I to produce fragments in the range of 0.5 kb to 1 kb (Sinclair et al. 1990). In this manner, a total of 50 probes were generated. Each of these probes was hybridized to Southern blots of DNA from humans (males and females), bovines (males and females), mice (males and females), and human-hamster somatic cell hybrids containing the human X or Y chromosome. All the probes were tested with and without prehybridization to total human DNA to suppress the effect of repetitive sequences. Despite this precaution, most probes did not detect unique sequences, but hybridized to repetitive elements distributed throughout the human and bovine genome.

Seven probes were found that detected single-copy Y-specific bands in human DNA. However, only one of these probes also gave Y-specific bands in the murine and bovine genomes. We named the putative gene defined by this probe *SRY* (*sex determining region-Y* gene). The equivalent mouse gene, *Sry*, is present in the smallest part of the mouse Y chromosome known to be sex determining, and *Sry* is deleted from a mutant Y chromosome that is no longer sex determining (Gubbay et al. 1990). Sequences homologous to *SRY* were present on the Y chromosome of all eutherian mammals tested. This degree of conservation suggests *SRY* has a functional role on the mammalian Y chromosome.

Problems encountered in the search for *TDF*

To identify sequences encoding transcripts within a genomic region, even in a small (35 kb) well-defined region such as ours, was difficult. Our approach of fragmenting the genomic DNA and using each fragment to probe "Noah's Ark" blots encountered some problems: (1) Actually generating the probes and making them small enough (less than 500 bp) is time-consuming and difficult. There is also the possibility that very small fragments may be lost during preparation. (2) The biggest obstacle we encountered was the number of repetitive elements detected by the probes in the genomic DNA of various species. To a degree, this could be reduced by prehybridization with total human DNA, but often this was of no help. Another possible approach we might have taken would be to search for HTF islands that are characterized by the presence of unmethylated CpG residues. Such CpG islands are often associated with genes and can be detected by digesting genomic DNA with restriction enzymes that recognize unmethylated CpG residues. However, in our case, this approach was of no use, because *SRY* is not associated with an island.

Perhaps the simplest and most obvious way to find transcripts is to use genomic fragments to probe mRNA on Northern blots. However, one problem is knowing in which tissue the gene is expressed. Even if this is known, then getting sufficient quantitites of the tissue for mass screening could be difficult. In our case, we expected the gene to be expressed in the fetal gonad at about 6 weeks. However, we could not obtain fetal gonads, and we used testes from elderly prostate cancer patients. Strangely, and luckily for us, *SRY* is expressed in adult testis mRNA, although we were not able to use this as a screen because it would have required mRNA from a large number of adult testes.

Two other approaches to isolating transcribed sequences are *Alu* PCR of human/rodent somatic cell hybrids and exon trapping (for a review of these methods and their limitations, see Hochgeschwender and Brennan 1991). However, these techniques are relatively new and are in the development stage. Both methods also require that the target gene has introns.

STEP 4: CORRELATIONS BETWEEN *SRY* AND *TDF*

The nucleotide sequence of *SRY* revealed two open reading frames of 99 and 223 amino acids, which overlapped in different frames. A search of the EMBL DNA sequence database failed to find any related sequences. Screening of the PIR protein database using the predicted proteins encoded by these open reading frames revealed homology to the longer open reading frame with several proteins. Homology was found with the Mc protein encoded by the *mat 3-M* gene of the fission yeast *Schizosaccharomyces pombe* (Kelly et al. 1988). The putative *SRY* protein also showed homology with a protein encoded by a gene at the *Neurospora crassa* a mating-type region (Staben and Yanfosky 1990), a conserved motif found in the nonhistone proteins, high mobility groups 1 and 2 (HMG1 and HMG2; Kolodrubetz 1991), the human nucleolar transcription factor hUBF (human upstream binding factor; Jantzen et al. 1990), and several transcription factors (Travis et al. 1991; van de Wetering et al. 1991). The common motif in all these proteins is a 79-amino-acid motif, known as the HMG box, that is associated with DNA-binding activity. Such a DNA-binding function for *SRY* would be consistent with a role in gene regulation.

The HMG box from several different proteins is presented in Figure 3. It is clear from this figure that the mouse autosomal genes (now called *Sox* genes), which were originally isolated by cross-hybridization with *SRY/Sry* sequences, form a related subgroup. Investigation of the expression patterns of three of the *Sox* genes suggests they have a role in neuronal differentiation (J. Collignon, pers. comm.). These genes are not expressed in the genital ridge.

In humans, *SRY* was shown to have a testis-specific transcript. In the fetal mouse, *Sry* is confined to the genital ridge, is germ-cell-independent and is expressed during 10.5–11.5 days postcoitum just prior to the first signs of testicular development (Koopman et al. 1990).

All the above-mentioned features are consistent with *SRY* having a role in developmental regulation of the testis and imply that *SRY* was an excellent candidate for *TDF*.

STEP 5: EQUATING *SRY* AND *TDF*
Analysis of *SRY* mutations in XY females

If *SRY* is required for testis formation, mutations in this gene should result in male to female sex reversal. This hypothesis could be tested by using homologous recombination to disrupt the *Sry* gene in embryonal stem cells. An alternative approach is to study XY females; a proportion of these individuals should have mutations in *SRY*.

The most comprehensive method for looking for mutations is to sequence the putative mutant gene and to compare it with the wild type; however, this can be very tedious and time-consuming. Several techniques have been developed that avoid the need to sequence large stretches of DNA. These include denaturing gradient gel electrophoresis (DGGE), RNase A cleavage, single-strand conformational polymorphism (SSCP) assay, and chemical cleavage. Initially, we chose to use the SSCP assay (Orita et al. 1989) to study *SRY* from XY females. This method exploits the ability of single-stranded DNA to form stable structures by intramolecular pairing and the separation of these structures by electrophoresis on native sequencing gels. The structures formed are very sensitive to small changes in sequences, and the separation is probably a function of the amount of secondary structure formed by the single strand of DNA. SSCP has the advantage of technical simplicity, but it may suffer the disadvantage of not detecting all possible mutations.

In the first set of experiments we investigated 11 XY females and 50 normal controls (Berta et al. 1990). The SSCP-polymerase chain reaction (PCR) products were codigested with two restriction enzymes to produce fragments in the optimal size range of 200 bp. DNA from 2 of

CLONING TDF

```
TCF-1       IKKPLNAFMLYMKEMRAKVIAECTLKESAAINQILGRRWHALSREEQAKYYELARKE    RQLHMQLYPGWSARDNYGKKKRR
                                           |||        |  |      |   |    |
Mc          ERTPRPPNAFILYRKEKHATLLKSNPSINNSQVSKLVGEMWRNESKEVRMRYFKMSEFYKAQHQKMYPGYKY    QPRKNKVK
                 ||||||   ||   |   |  |     |||| |||||    ||         ||| |   |  ||     ||
Mouse SRY   GHVKRPMNAFMVWSRGERHKLAQQNPSMQNTEISKQLGCRWKSLTEAEKRPFFQEAQRLKILHREKYPNYKY    QP  HR    RAKV
            ||||||||||||||||||  || | || | |||||| |||||||||||||||||||| |||| |||||||||    ||  ||
Human SRY   DRVKRPMNAFIVWSRDQRRKMALENPRMRNSEISKQLGYQWKMLTEAEKPFFQEAQKLQAMHREKYPNYKY     RPR  RK    AKM
            ||||||||| |||| |||| || || ||||||||| || ||| || |||||||| | |  |||| ||||
Mouse a4    PSGHIKRPMNAFMVWSQIERRKIMEQSPDMHNAEISKRLGKRWKLLKDSDKIPFIQEAERLRLKHMADYPDYKY    RPR  KK    VKSG
                 ||||||||||   ||||    |    ||||| |||| |         |||| |||| ||    ||||
Mouse a1    NQDRVKRPMNAFMVWSRGQRRKMAQENPKMHNSEISKRLGAEWKVMSEAEKRPFIDEAKRLRALHMKEHPDYKY    RPRR KT    K
            ||||||||||||||||| |||||||||||||||||||||||||| | ||||||||||||||||| |||||||
Mouse a2    SPDRVKRPMNAFMVWSRGQRRKMAQENPKMHNSEISKRLGAEWKLLSETEKRPFIDEAKRLRALHMKEHPDYKY    RPRRKTKTLMKKDKYTL
            |||||||||||||||||||||||||||||||||||||||  |  ||||||||||||| ||||||||||
Mouse a3    DQDRVKRPMNAFMVWSRGQRRKMALENPKMHNSEISKRLGADWKLLTDAEKRPFIDEAKRLRAVHMKEYPDYKY    RPRRKTKTLLKKDKYSL
                                  |||                         |                    ||
HMG-1       GKGDPKKP RGKM  SSYAFFVQTCREEHKKKHPDASVNFSEFSKKCSERWKTMSAKE K GKFEDMAKADKARYER  EMKTYIPPKGETKKK
                      ||
HUBF-1      kK LKKHPDFPK KPLTPYFRFFMEKRAKYAKLHPEMSNLDLTKILSKKYKELPEK KK MKYIQDFQREKQEFERNLARFREDHPDLIQNAK
```

Figure 3 Sequence comparisons of proteins related to SRY. The HMG "boxes" of several proteins aligned using the method of Needleham and Wunsch (1970). The lines represent identities between adjacent sequences; conservative amino acid changes are not indicated.

the 11 patients gave abnormal SSCP patterns for *SRY*. The father of one (AA) gave a normal pattern and the father of the other (JN) gave an abnormal pattern. Further PCR amplifications were performed on these two patients and their fathers. The PCR products were purified, end-repaired, and cloned by blunt-end ligation into plasmids. The cloned PCR products were then sequenced by conventional plasmid DNA sequencing techniques. The sequencing data supported the SSCP data: AA had a de novo mutation and JN had an inherited variant. The results from AA indicate that *SRY* is required for testis formation, supporting the hypothesis that *SRY* is *TDF*. A similar conclusion was reached by Jager et al. (1990). It is not known if the inherited variant in the father and daughter (JN) is capable of causing sex reversal.

Subsequently, the study of *SRY* mutations in XY females was extended to a further 23 patients. This time they were subjected not only to SSCP analysis, but also to sequencing of the gene. The purpose of employing both methodologies was to establish the efficiency of SSCP in detecting point mutations.

To make sequencing 800 bp from 23 individuals (18 kb in total) less tedious, *Eco*RI sites were incorporated into the PCR primers with an 8-nucleotide tail to facilitate *Eco*RI digestion. The PCR products were digested directly in the PCR buffer and cloned into plasmids. This was more efficient than the cloning method employed before. A single clone from each patient was sequenced. Sequencing gels were not run in the conventional manner, but a single termination reaction from each clone (i.e., individual) was run on the gel together. This did not give a sequence as such but gave a pattern that did not need to be read, in which mutations were present as the loss or gain of a band. From the 23 clones, 8 sequence differences were found (Fig. 4).

The SSCP analysis of these 23 individuals employed the same restriction digests as before, but included two additional sets of digests, in case mutations near the ends of restriction fragments had gone undetected. Of the 23 samples, only 3 gave abnormal SSCP patterns.

Because some, or all, of the mutations found by sequencing could have been PCR artifacts, i.e., *Taq* polymerase errors introduced during amplification, each potential mutation was subjected to direct sequencing of the PCR products. Only five of the eight "mutations" were shown to be genuine, the other three being PCR artifacts. Of the five genuine mutations, three were SSCP-positive. One of the SSCP-negative mutation-positive samples was subjected to a further restriction enzyme digest for SSCP analysis and this time showed a very subtle bandshift. Thus, in total, four of the five mutations were detected by SSCP. Whether the difficulty in detecting all of the mutations by SSCP reflects the limitations of the technique or our inability to conduct the experiments properly is unclear.

Three PCR artifacts were observed in 18 kb of sequenced DNA.

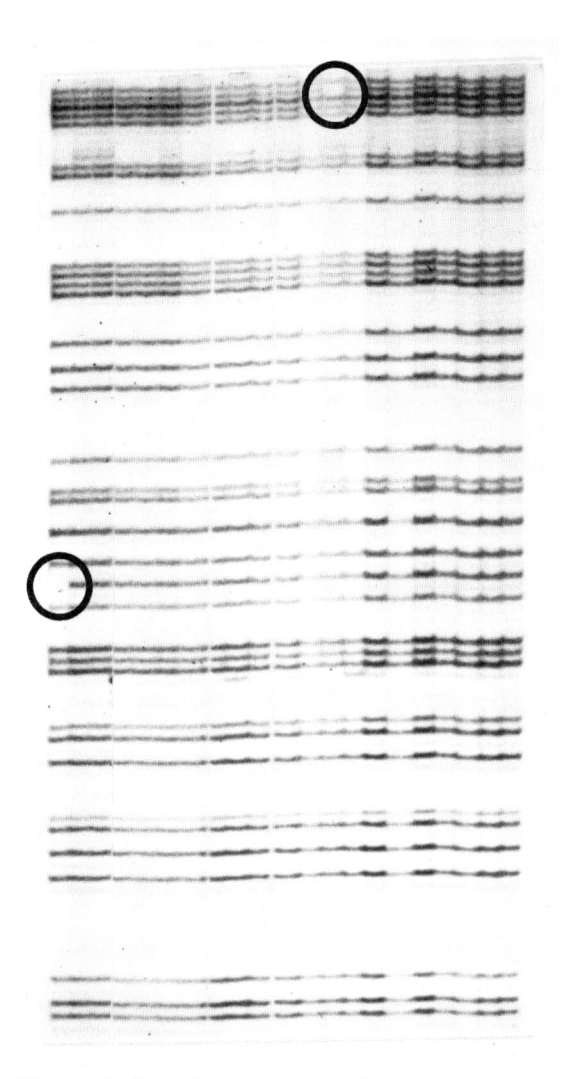

Figure 4 A-track sequencing of *SRY* from XY females. The figure shows a portion of a single termination-reaction sequencing gel, used for the rapid identification of mutations in *SRY*. The gel illustrates mutations in 2 of 21 clones (individuals). The mutation in clone 1 is an A to G transition and that in clone 13 is an A to T transversion. Both are seen on this "ddA gel" as the loss of a band and are marked with circles.

This rate is in keeping with estimates of *Taq* polymerase infidelity measured by others. It is apparent that high cycle numbers and low extension temperatures contribute to the error rate. It is therefore advisable to determine the minimum number of cycles and the maximum extension temperature for a particular pair of primers. Alternative sequencing approaches could be used to avoid the problem of PCR artifacts. One possibility would be orientation-specific cloning of the PCR products and mixing together several different clones from each individual. Another approach is to sequence the PCR products directly. In conclusion, SSCP is a powerful and simple techique, but it may only detect about 75–80% of point mutations.

Construction of transgenic mice

The analysis of XY females cannot resolve whether *SRY* is the only gene on the Y chromosome required for male sex determination. This can only be resolved by injecting *Sry* sequences into XX mouse embryos. If *Sry* is the only Y-located gene needed for male sex determination, the fate of the embryos should be changed from female to male.

Lovell-Badge and colleagues (Koopman et al. 1991) injected mouse embryos with a 14-kb fragment of the mouse Y chromosome that included *Sry*. The injected fragment is considerably larger than the known transcribed sequences; this was delibertately done because the *Sry* gene is very tightly regulated in development, and it was hoped that a larger fragment might include all the sequences needed for correct regulation of the gene. One in four XX embryos that incorporated the *Sry* transgene was sex-reversed. One of these embryos was allowed to go to term; the male animal produced had small testes and normal internal and external genitalia. His mating behavior was also normal. The small testes are due to the failure of spermatogenesis—this failure is always seen in male animals with two X chromosomes, even in the presence of a complete Y chromosome. It is also possible that the Y chromosome encodes one or more genes required for spermatogenesis (Burgoyne 1987). Position effects on the expression of the transgene could explain those XX animals that incorporated *Sry* sequences but failed to develop as males.

The transgenic experiment proves that the 14-kb fragment contains all the Y-located sequences required for male sex determination in an XX embryo. This fragment has been sequenced, and *Sry* is the only gene that has been found.

CONCLUSIONS

Biological conclusions

SRY is the Y-located male sex-determining gene. This conclusion is based on the following:

1. *SRY/Sry* maps to the smallest region of the Y chromosome known to be sex-determining in man and mouse. Homologous

sequences are found on the Y chromosomes of all eutherian mammals tested.
2. *Sry* is expressed at the right time and right place for a mammalian sex-determining gene.
3. Mutations in *SRY* can cause female to male sex reversal.
4. *Sry* sequences can cause male to female sex reversal in transgenic mice.

Genomic conclusions

In cloning *TDF*, the major problems were in chromosome walking and identification of genes. Advances in technology have greatly facilitated chromosome walking. Gene identification is still difficult—even given a complete sequence, it is not always possible to exclude the presence of a gene.

Acknowledgments

We are grateful to all our friends in the Laboratory of Human Molecular Genetics who have contributed blood, sweat, and many tears to this work. All our recent work on sex determination has been in collaboration with Robin Lovell-Badge and his co-workers. This manuscript was prepared with the help of Clare Middlemiss.

References

Berta, P., J.R. Hawkins, A.H. Sinclair, A. Taylor, B. Griffiths, P.N. Goodfellow, and M. Fellous. 1990. Genetic evidence equating *SRY* and the testis-determining factor. *Nature* 348: 448.

Brown, W.R.A. 1988. A physical map of the human pseudoautsomal region. *EMBO J.* 7: 2377.

Burgoyne, P.S. 1982. Genetic homology and crossing over in the X and Y chromosomes of mammals. *Hum.Genet.* 61: 85.

———. 1987. The role of the mammalian Y chromosome in spermatogenesis. *Development* 101S: 133.

Burke, D.T., G.F. Carle, and M.V. Olson. 1987. Cloning of large segments of exogenous DNA into yeast by means of artificial chromosome vectors. *Science* 236: 806.

Cooke, H.J., W.R.A. Brown, and G. Rappold. 1985. Hypervariable telomeric sequences from the human sex chromosomes are pseudoautosomal. *Nature* 317: 688.

Ellis, N.A., P.J. Goodfellow, B. Pym, M. Smith, M. Palmer, A.-M. Frischauf, and P.N. Goodfellow. 1989. The pseudoautosomal boundary in man is defined by an *Alu* repeat sequence inserted on the y chromosome. *Nature* 337: 81.

Ferguson-Smith, M.A. 1966. X-Y chromosomal interchange in the aetiology of

true hermaphroditism and of XX Klinefelter's Syndrome. *Lancet* **II:** 475.
Ford, C.W., K.W. Jones, P. Polani, J.C. De Almedia, and J.H. Brigg. 1959. A sex chromosome anomaly in a case of gonadal sex dysgenesis (Turner's Syndrome). *Lancet* **I:** 711.
Goodfellow, P.J. and S.M. Darling. 1988. Genetics of sex determination in man and mouse. *Development* **102:** 251.
Goodfellow, P.N., S.M. Darling, and J. Wolfe. 1985. The human Y chromosome. *J. Med. Genet.* **22:** 332.
Goodfellow, P.J., S.M. Darling, N.S. Thomas, and P.N. Goodfellow. 1986. A pseudoautosomal gene in man. *Science* **234:** 740.
Goodfellow, P., G. Banting, D. Sheer, H.H. Ropers, A. Caine, M.A. Ferguson-Smith, S. Povey, and R. Voss. 1983. Genetic evidence that a Y-linked gene in man is homologous to a gene on the X chromosome. *Nature* **302:** 346.
Gubbay, J., J. Collignon, P. Koopman, B. Capel, A. Economou, A. Munsterberg, N. Vivian, P.N. Goodfellow, and R. Lovell-Badge. 1990. A gene mapping to the sex-determining region of the mouse Y chromosome is a member of a novel family of embryonically expressed genes. *Nature* **346:** 245.
Guellaen, G., M. Casanova, C. Bishop, D. Geldwerth, G. Audre, M. Fellous, and J. Weissenbach. 1984. Human XX males with Y single-copy DNA fragments. *Nature* **307:** 172.
Hochgeschwender, U. and M.B. Brennen. 1991. Identifying genes within the genome: New ways for finding the needle in a haystack. *BioEssays* **13:** 139.
Jacobs, P.A. and J.A. Strong. 1959. A case of human intersexuality having a possible XXY sex-determining mechanism. *Nature* **183:** 302.
Jager, R.J., M. Anvret, K. Hall, and G. Scherer. 1990. A human XY female with a frameshift mutation in *SRY*, a candidate testis-determining gene. *Nature* **348:** 452.
Jantzen, H.M., A. Admon, S.P. Bell, and R. Tjian. 1990. Nucleolar transcription factor *hUBF* contains a DNA-binding motif with homology to HMG proteins. *Nature* **344:** 830.
Jost, A., B. Vigier, J. Prepin, and J.P. Perchellet. 1973. Studies on sex differentiation in mammals. *Recent Prog. Horm. Res.* **29:** 1.
Kelly, M., J. Burke, M. Smith, A. Klar, and D. Beach. 1988. Four mating-type genes control sexual differentiation in the fission yeast. *EMBO J.* **7:** 1537.
Koller, P.C. and C.D. Darlington. 1934. The genetical and mechanical properties of the sex chromosomes. 1. *Rattus norvegicus*. *J. Genet.* **29:** 159.
Kolodrubetz, D. 1990. Consensus sequence for HMG1-like DNA binding domains. *Nucleic Acids Res.* **18:** 5565.
Koopman, P., J. Gubbay, N. Vivian, P. Goodfellow, and R. Lovell-Badge. 1991. Male development of chromosomally female mice transgenic for *Sry*. *Nature* **351:** 117.
Koopman, P., A. Munsterberg, B. Capel, N. Vivian, and R. Lovell-Badge. 1990. Expression of a candidate sex-determining gene during mouse testis differentiation. *Nature* **348:** 450.
Levilliers, J., B. Quack, J. Weissenbach, and C. Petit. 1989. Exchange of terminal portions of the X and Y chromosome short arms in human XY females. *Proc. Natl. Acad. Sci.* **86:** 2296.
Needleman, S.B. and C.D. Wunsch. 1970. A general method applicable to the search for similarities in the amino acid sequence of two proteins. *J. Mol. Biol.* **48:** 443.

Orita, M., Y. Suzuki, T. Sekiya, and K. Hayashi. 1989. Rapid and sensitive detection of point mutations and DNA polymorphisms using the polymerase chain reaction. *Genomics* **5:** 874.

Page, D.C., E.M.C. Fisher, B. McGillivray, and L.G. Brown. 1990. Additional deletion in sex-determining region of human Y chromosome resolves paradox of X,t(Y;22) female. *Nature* **346:** 279.

Page, D.C., R. Mosher, E.M. Simpson, E.M.C. Fisher, G. Mardon, J. Pollack, B. McGillivray, A. de la Chapelle, and L.G. Brown. 1987. The sex-determining region of the human Y chromosome encodes a finger protein. *Cell* **51:** 1091.

Palmer, M.S., A.H. Sinclair, P. Berta, N.A. Ellis, P.N. Goodfellow, N.E. Abbas, and M. Fellous. 1989. Genetic evidence that *ZFY* is not the testis-determining factor. *Nature* **342:** 937.

Petit, C., J. Levilliers, and J. Weissenbach. 1988. Physical mapping of the human pseudoautosomal region; comparison with genetic linkage map. *EMBO J.* **7:** 2369.

Pritchard, C.A., P.J. Goodfellow, and P.N. Goodfellow. 1987. Mapping of the limits of the human pseudoautosomal region and a candidate sequence for the male-determining gene. *Nature* **328:** 273.

Rouyer, F., M.C. Simmler, C. Johnsson, G. Vergnaud, H.J. Cooke, and J. Weissenbach. 1986. A gradient of sex-linkage in the pseudoautosomal region of the human sex chromosomes. *Nature* **319:** 291.

Schneider-Gadicke, A., P. Beer-Romero, L.G. Brown, R. Nussbaum, and D.C. Page. 1988. *ZFX* has a gene structure similar to *ZFY*, the putative human sex determinant and escapes X-inactivation. *Cell* **57:** 1247.

Sinclair, A.H., J.W. Foster, J.A. Spencer, D.C. Page, M. Palmer, P.N. Goodfellow, and J.A.M. Graves. 1988. Sequences homologous to *ZFY*, a candidate for the sex-determining gene, are autosomal in marsupials. *Nature* **336:** 780.

Sinclair, A.H., P. Berta, M.S. Palmer, J.R. Hawkins, B.L. Griffiths, M.J. Smith, J.W. Foster, A.M. Frischauf, R. Lovell-Badge, and P.N. Goodfellow. 1990. A gene from the human sex determining region encodes a protein with homology to a conserved DNA-binding motif. *Nature* **346:** 240.

Staben, C. and C. Yanofsky. 1990. *Neurospora crassa a* mating-type region. *Proc. Natl. Acad. Sci.* **87:** 4917.

Travis, A., A. Amsterdam, C. Belanger, and R. Grosschedl. 1991. LEF-1, a gene encoding a lymphoid specific protein with an HMG domain, regulates T-cell receptor and enhancer function. *Genes Dev.* **5:** 880.

van de Wetering, M., M. Oosterwegel, D. Dooijes, and H. Clevers. 1991. Identification and cloning of *TCF-1*, a T lymphocyte-specific transcription factor containing a sequence-specific HMG box. *EMBO J.* **10:** 123.

Vergnaud, G., D.C. Page, M.C. Simmler, L. Brown, F. Rouyer, B. Noel, D. Botstein, A. de la Chappelle, and J. Weissenbach. 1986. A deletion map of the human y chromosome based on DNA hybridisation. *Am. J. Hum. Genet.* **38:** 172.

Weissenbach, J., J. Levilliers, C. Petit, F. Rouyer, and M.C. Simmler. 1987. Normal and abnormal interchanges between the human X and Y chromosomes. *Development* **101s:** 67.

Genetic Analysis of Multifactorial Disease: Lessons from Type-1 Diabetes

Soumitra Ghosh and John A. Todd

Nuffield Department of Surgery
John Radcliffe Hospital
Headington, Oxford OX 3 9DU
United Kingdom

The genetics of complex diseases has come to the forefront of analytical study. After the successful identification of genes for monogenic disorders such as cystic fibrosis, research is focusing on attaining similar goals in diseases with non-Mendelian inheritance, such as insulin-dependent diabetes mellitus (IDDM), schizophrenia, Alzheimer's disease, hypertension, and atherosclerosis. The essential feature of these disorders is that both environmental and genetic influences contribute to disease onset and progression.

The aims of this chapter are:

❏ to provide a concise review of the methodology used in the analysis of complex disease genetics

❏ to highlight IDDM as a good example for analysis and show the strength of studying animal models to dissect out the greater complexities that exist in humans

INTRODUCTION

Multifactorial diseases such as those described above are common and cause major health care problems in the Western world. Preventing these illnesses in subjects with a genetic predisposition will require knowledge of the environmental triggering factors; identification of the genes is the first step toward prevention.

Many of these complex diseases have quantitative biological correlates that can be measured (metric characters), e.g., serum transferrin levels in hemochromatosis. Blood pressure in the population, for example, follows an approximately normal distribution, but when it exceeds a certain threshold, hypertension is diagnosed. This artificial cutoff level defines hypertension as a discrete trait rather than as the extreme end of a normal distribution. However, the severity of the complications of raised blood pressure (e.g., renal failure and cerebral atherosclerosis) is proportional to its level, and a certain amount of end-organ damage will occur even in those individuals deemed to have normal blood pressure. Therefore, underlying a discrete disease may be a continuous (quantitative) process; the medical importance lies in its absolute level as well as when the level exceeds a certain threshold.

Earlier in the century, biometricians were able to define close likeness among quantitative traits in families, suggesting a genetic mechanism (although this could have been due partly to common environmental sharing). However, since these conditions do not show classic Mendelian segregation patterns, inheritance of particulate matter (genes) underlying these phenomena was questioned.

Fortunately, the two schools of genetics, Mendelian and biometric, converged with the recognition of the existence of polygenes. These are multiple genes, each with small effect, but which add up to a measurable quantitative change. The phenotype is further modified by the environment, giving rise to continuous characteristics that approximate to a normal distribution. Nonadditivity, or interactions between genes (epistasis), and intralocus interactions in the form of dominance relationships between alleles may cause skewing of this distribution. In this paper, we use epistasis interchangeably with multiplicative, although the latter is a specific subset of the former.

We present a recent update on analysis of complex diseases and also discuss methods for tackling the arising problems in IDDM. Other aspects of recent advances in genetic mapping are discussed by Lander (1988) and Risch (1991b). The major histocompatibility complex (MHC) is not discussed in detail in this paper; for recent articles, see Leigh-Field (1988), Thompson (1988), Todd (1990a,b), and Tait and Harrison (1991).

APPROACHES TO COMPLEX DISEASE STUDIES
Likelihood

The concept of likelihood has many applications in genetic statistics. Given a hypothesis H and data D, likelihood is formally defined as the probability of the hypothesis given the data, $Pr\ (H/D)$, for some specified model (Edwards 1972). For example, a bag contains a large number of balls, half of which are red and half black. If one picks out three balls and finds two black and one red, probability tells us that this event occurs with chance 3 x (1/2 x 1/2 x 1/2) = 3/8, regardless of order, provided the balls are replaced each time.

Likelihood, however, gives us different values, depending on the hypotheses (if we had no idea that in reality the ratio of black to red was 1:1). For a hypothesis of 2:1, it will be 3 x (2/3 x 2/3 x 1/3) = 4/9, which is greater than 3/8, the likelihood for a hypothesis of 1:1. Thus, likelihood states that with the limited data, the population ratio is more likely to be 2:1 than 1:1 by odds of 4/9 x 8/3 = 1.19:1. The larger the sample, the more apt the likelihood estimate is to give the real population value.

Maximum likelihood is used as an estimation procedure for a variable in the likelihood equation. If in a backcross mating AaBb x aabb there are four offspring of which three are nonrecombinants, AaBb, and one is recombinant, Aabb, the likelihood L, as a function of the recombination estimate θ, is $L(\theta) = \theta (1 - \theta)^3$. Now the maximum likelihood estimate (MLE) of θ is found by maximizing the above equation with respect to θ. An estimate of $\theta = 1/4$ gives the maximum value of $L(\theta)$, 27/64. This is intuitively appealing, as the most logical estimate for θ, given the data, would have been 1/4 (one recombinant out of four). With more complicated likelihood equations involving pedigrees and missing information (when likelihood can be used to give weights to the possibilities), maximization of $L(\theta)$, even by using calculus, becomes impossible. Algorithms can then be used to arrive at a close approximation (e.g., EM algorithm) in these situations.

The likelihood ratio (LR) statistic is used for hypothesis testing for the MLE compared to the null hypothesis estimate. To compare with the hypothesis of no linkage in our hypothetical family, we have

$$L(\theta = 1/4)/L(\theta = 1/2) = \frac{1/4 (1-1/4)^3}{1/2 (1-1/2)^3}$$
$$= 27/16 = 1.68$$

Log_{10} of this result gives the LOD score for this family, 0.23. Depending on the model and its constraints (which will determine the degrees of freedom)

$$-2\log_e L (\theta_0)/L (\theta_1)$$

with large numbers will have an approximate χ^2 distribution. θ_0 represents the MLE under the null hypothesis; θ_1 represents the MLE under the alternative hypothesis. Thus, we can test the significance of the likelihood ratio statistic by use of χ^2 tables. Taking a likelihood ratio of 1000 gives a LOD score of 3. This is equivalent to $-2\log_e 1/1000 = 13.8$ on the χ^2 distribution scale, provided regularity conditions hold. Support is defined as the natural logarithm of the likelihood ratio. Other variables, such as penetrance and gene frequency, can appear in the likelihood equation.

Penetrance and phenocopies

Penetrance is defined as the proportion of the population with a susceptible genotype that shows the expected phenotype. The effect of reduced penetrance when full penetrance is assumed is to cause inconsistency in the recombination estimate (Ott 1985). Unaffecteds with the disease genotype will overestimate θ, and linkage may be missed. Ignoring unaffecteds in the analysis will increase the power to detect linkage (Hyer et al. 1991; Pericak-Vance et al. 1991).

Age-dependent penetrance is apparent in multifactorial disease appearing later in life. One can either make a number of discrete estimations from a cumulative frequency graph for the disorder or make penetrance a continuous function of age. An elegant method is via the logistic function (Bonney 1986); this type of modeling can help improve support intervals for disease location (Todd et al. 1991).

Phenocopies, for example, noninsulin-dependent diabetics (NIDDM) in a supposedly IDDM population, will increase the chance of rejecting true linkage (β error). This and other features of sample selection (ascertainment) are important prerequisites for genetic studies.

Genetic heterogeneity

If a disease can be caused by one of two or more unlinked genes and significant epistasis is not acting, all families analyzed together may not show linkage to any locus. Various methods have been put forward to test for genetic heterogeneity, and estimating for a proportion of all families linked to one of two loci has been the most common solution (Smith 1963). The likelihood equation then has two parameters: a, the number of families linked to a locus, and θ. Using HOMOG 1, MLEs can be found for both these variables.

Heterogeneity in tuberous sclerosis, with linkages to both chromosomes 9 and 11, has been analyzed by constructing a hypothetical hybrid chromosome (Janssen et al. 1990). Maps of the two chromosomes were oriented in a head-to-tail fashion and the junction was assumed not to be linked. LOD scores using multipoint linkage analysis were calculated for the adjacent regions. HOMOG 2 is able to analyze families showing linkage to two separate areas of the same chromosome; this program was used on the data from the synthetic chromosome. Heterogeneity was favored with one disease locus on either side of the imaginary junction. This method is more sensitive than the standard method of Smith (1963) for detecting heterogeneity.

Lander and Botstein (1986) have advocated interval mapping and simultaneous search procedures incorporated into MAPMAKER (Lander et al. 1987) to surmount heterogeneity. Interval mapping uses likelihood ratios comparing the likelihood for a disease gene's being in an interval between the two markers with the likelihood of its being outside this interval. The expected contribution to the LOD score is greatest when there is no recombination in the interval containing the disease locus. Recombination between disease locus and flanking markers contributes little. Therefore, families are linked to a locus either tightly or not at all. Simultaneous search helps dissect out heterogeneity by looking at many loci together. The likelihood of the trait cosegregating with any one of the loci is compared to no segregation at all by a LR test. Multiple hypotheses can be studied to analyze even more complex multilocus models.

Linkage disequilibrium

Alleles of two loci are said to be in equilibrium when the gametic frequencies for all possible haplotypes equals the product of individual allele frequencies. Disequilibrium exists when this is not so and may occur with linked or unlinked genes. The measure of disequilibrium is taken to be half the difference in frequency between coupling and repulsion heterozygotes (Falconer 1989). Specifically, disequilibrium disappears more quickly for unlinked loci and with increasing generations (more meiotic events). Thus, after t generations, disequilibrium D_t is given by $D_t = D_0 (1-\theta)^t$, where D_0 denotes disequilibrium at generation 0.

Linkage disequilibrium is most apparent with HLA (human leukocyte antigen) and disease associations (e.g., in IDDM, a strong association with A1B8DR3 exists). The genesis of haplotypic association is now fully appreciated when one considers the reduced recombination present in the HLA region.

Allelic association is an uncertain measure of genetic distance, yet for cystic fibrosis, this was a major step for delineating the whereabouts of the gene (Kerem et al. 1989). Although there were fluctuations in disequilibrium, researchers were able to narrow the area of the major mutation to about 100 kb from the limit of genetic mapping of approximately 900 kb. This is a major method to locate genes and is applicable to complex diseases, as demonstrated by several well-established associations of HLA and diseases.

Epistasis confounding linkage disequilibrium

The identity by descent (ibd) method of affected sib pair (ASP) analysis relies on the fact that two sibs will share any allele(s) unlinked to disease in Mendelian proportions of 1:2:1 for 2, 1, or 0 alleles, respectively. If the marker allele is linked to the disease gene, then sharing of 2 alleles ibd increases and, likewise, the sharing of 0 alleles decreases. The extent of this distortion can be compared to expected proportions by likelihood ratios, which give an idea about the strength of linkage. The method is commonly applied to complex diseases because it makes no assumptions on the mode of inheritance. It has been most extensively used for the HLA region.

One plausible explanation for marker/trait association is that the disease locus is tightly linked and in linkage disequilibrium with the marker locus (Hodge 1981). Another explanation is that the marker may be implicated in disease etiology in epistatic interaction with another unlinked disease locus (Clerget-Darpoux and Bonaiti-Pellie 1980). This phenomenon would explain both the disease marker association and the increased haplotype sharing in families for HLA. Genes of the HLA system could be said to have pleiotropic effects, as there may be many disease/marker associations. Thus, ibd distributions and association data would not distinguish between linkage disequilibrium and pleiotropy.

Risch (1987) has partly solved this problem by utilizing a sib-discordance method: For a single locus, he derives the expression for the posterior probability (p_i) for sibs sharing i alleles ibd (where i = 0, 1, 2), given that they are both affected. This turns out, for example, for i = 0, to be equal to the prior probability of sibs sharing 0 alleles ibd (= 0.25) divided by λ_s (the risk of disease for sibs of a proband [K_s] divided by population prevalence [K]). This method assumes no recombination between marker and disease. K_s may be estimated in two ways: by the above method utilizing observed sib-discordance data or by the method of James (1971). James's formula relates K_s to K for a discrete trait in terms of covariance (this measures the degree of resemblance between relatives for a trait) and penetrance from population and disease samples. If this value of K_s is below that for the sib-discordance method, linkage disequilibrium is favored over pleiotropy. This is because if alleles are in linkage disequilibrium for marker and disease loci, sib pair discordance will be very similar for both loci. Therefore, K_s will be falsely overestimated for the marker locus, e.g., HLA compared to the case of no disequilibrium. The true value given by the James formula does not suffer from this distortion and will be less than the value from the sib-discordance method for linkage disequilibrium, but equal to it in the presence of pleiotropy. For IDDM, and DR3/4, K_s (James) = 1.77 and K_s (Risch) = 3.42, favoring linkage disequilibrium over pleiotropy.

Complex segregation and pedigree analyses

Segregation analysis will determine by mathematical modeling the mode of inheritance of a particular phenotype among randomly sampled families (Elston 1981). These may be nuclear or extended pedigrees, for which ascertainment schemes are less well defined. Modes of inheritance are determined by transmission probabilities, π (chance of transmitting an allele to an offspring given genotype).

Classic segregation analysis was limited to discrete phenotypes. Modifications were needed for quantitative traits. Complex segregation analysis subsumes the effects of a major locus or loci, polygenes, and environment. Lalouel et al. (1983) have proposed a unified model amalgamating older methods. They suggest that a trait results from an inheritable component, a multifactorial transmissible part, and a random environmental component. Such a system has been incorporated into the package POINTER. Various hypotheses can be formulated, for example, by comparing observed and expected π for different models. Estimation and hypothesis testing are by likelihood methods. The unified model allows a more comprehensive investigation about the existence and nature of family transmission. Other parameters estimated for the disease include penetrance, dominance relationships, multilocus, and environmental effects.

Segregation analyses performed with four early studies in IDDM and HLA (Risch 1984) show a rejection of recessive inheritance but not dominant inheritance. Although association with DR3 and DR4 was strong, the presence of other genetic influences not linked to HLA was suggested. It is more useful to do linkage and segregation analyses together to identify the parameters involved concomitantly. Certainly one must have enough evidence for a genetic component before beginning a study, or at least evidence for a possible candidate gene (Morton 1990).

Computational sophistication is a major requirement in complex segregation analysis, especially for large pedigrees. Schork (1991) has used vector and parallel processor computers in evaluating genetic models. Each processor will compute log likelihoods for assigned families, and vectorization will allow simultaneous genotype-phenotype relationships to be evaluated within each family. Computation time can be reduced considerably, and more pedigrees can be handled in an analysis.

Having established that a disease is segregating at a major locus or loci, mapping becomes the next priority. Boehnke et al. (1990) have devised, for quantitative traits, a new method to select pedigrees most likely to give the maximum information. This will reduce the cost of a study as well as provide a more efficient and less labor-intensive protocol. The method measures the expected number of potentially informative meioses (EIM) for each pedigree and is useful for disease loci with common alleles. The calculations are based on a Bayesian equation to find the number of informative meioses in a pedigree, given the quantitative trait information.

The group compares this method with another method based on a likelihood ratio statistic for pedigrees. This looks at the likelihood of a pedigree under a model allowing for a major locus compared to the likelihood for its exclusion. The conclusion is that the EIM method is sometimes more powerful in aiding selection of informative pedigrees for linkage studies.

Comparative mouse-human gene mapping

An increasingly powerful method for mapping complex disease genes is the use of an animal model (Rise et al. 1991; Todd et al. 1991). Conserved syntenies between mouse and human genomes are becoming more apparent as mapping studies progress (Nadeau 1989; Erickson 1990). Conserved linkages where both gene order and synteny are maintained are not so clear-cut. However, the two species are apparently near to saturation for homology segments and provide an excellent way to study complex disease. This has been proved already for the dysmorphic mouse syndromes. The mutation that gives rise to Waardenberg's syndrome has been mapped to 2q (Foy et al. 1990). The homologous region in mouse is on chromosome 1, and the splotch mutation (giving rise to pigmentary changes and deafness as in Waardenberg's syndrome) resides here.

For a complex trait, IDDM provides an example of mouse-human genetic homology. Alleles of the human HLA-DQBI locus, DQβ, have been sequenced, and a non-aspartic acid (alanine, valine, or serine) residue at position 57 has been shown to be correlated with IDDM in many cases (Todd et al. 1987). The equivalent molecule in mouse coding for I-Aβ has serine in position 57 for NOD (non-obese diabetic; see below) mice and an aspartic acid residue in diabetic-resistant strains such as NON (non-obese normal) (Acha-Orbea and McDevitt 1987). This residue probably has an important role in IDDM and is conserved across human and mouse.

Using an F2 rat cross, two recent studies show the successful mapping of loci important in hypertension (Hilbert et al. 1991; Jacob et al. 1991). One locus is strongly linked to angiotensin converting enzyme (chromosome 10), an obvious candidate gene. Both groups use standard quantitative genetic methods and estimate the percentage of total genetic variance explained by these loci.

Linkage analysis in complex traits

Risch (1990a,b,c) has made significant advances into the theoretical aspects of linkage analysis in complex traits. The results are for binary conditions: affected or unaffected. He utilizes λ_R, the risk of relatives of a proband versus population prevalence, and shows it to vary with the degree of relationship in different inheritance models. For example, λ_R-1 decreases by a factor of 2 for each degree of relationship (e.g., sibs [1st], uncle-niece [2nd], cousins [3rd]) in monogenic traits, and the same is true for additive models (locus-specific penetrances act additively). With epistatic models (locus-specific penetrances act multiplicatively), the drop is more pronounced. IDDM is shown by this method to follow an epistatic model more closely. Risch also shows how the investigator can test the hypothesis of the number of genes involved in a complex disease and how these genes act together. The ability to detect a gene depends on the contribution it makes to the variation of the trait.

The difference between the posterior probability of affected relative pairs sharing i alleles ibd and Mendelian expected values, utilizing the ASP method, gives a deviation parameter to determine power to detect linkage in terms of λ_R and θ (between marker and disease locus). This deviation parameter is next related to multiplicative and additive models. It is shown that comparative power to detect linkage in different relative pairs is determined by locus-specific λ_R for multiplicative models. For additive models, it is the total λ_R that determines comparative power. This has major implications: It is much easier with closer relative pairs to detect an individual locus in diseases with epistatic acting loci than to detect a locus in diseases with additively acting loci. Additive loci are just as easy to detect by using distant relative pairs. Risch (1990a,b,c) makes it clear, with examples, that his conclusions are dependent on specific λ_R and θ values.

For IDDM, the observed HLA discordance frequency data for uncle (aunt)-nephew (niece) pairs (p_0 = 0.21) is very similar to that expected for a locus in a multiplicative model. This result is based on very small numbers and needs to be reconfirmed (5/24 pairs). Cousin pairs would give the greatest power to detect linkage based on the deviation parameter for a large genetic study into IDDM, regardless of the types of interlocus relationships. Multipoint analysis is shown to improve precision of disease locus location and to improve power when 100% polymorphic markers are not used.

With increasing θ, power to detect linkage drops drastically, but now distant relative pairs give much greater power than sib pairs. Finally, Risch describes a maximum LOD statistic for affected relative pairs. This makes maximum use of uninformative families and gives MLEs for p_0, p_1, and p_2. By making this statistic a function of PIC (polymorphism information content of marker) and λ_R, an expected (average) maximal LOD score (MLS) is calculated for samples with differing numbers of sib pairs.

Two sampling strategies, one with just relative pairs and a second with information from additional relatives, are compared. Low PIC dramatically reduces maximal and expected LOD scores. This effect is different with different relative pairs, but power (=proportion of 5000 simulated replicate samples with MLS>3) is always greater with the second strategy. Risch concludes that for sib pairs and PIC < 0.7, the second scheme is more effective (in terms of expected LOD score per person typed). In summary, with the advent of more polymorphic markers such as microsatellites, sib pair analysis without typing additional relatives can be justified.

Polymorphisms

Identification of restriction-fragment-length polymorphisms (RFLPs) and its application to gene mapping (Botstein et al. 1980) paved the way for further developments, such as the discovery of minisatellites and application of these to genetic mapping (Nakamura et al. 1987). Minisatellites usually have more than two alleles, unlike RFLPs, and have greatly accelerated gene mapping in the last few years.

Microsatellites represent the third generation of marker loci. They are short, tandemly repeated sequences easily amplified by polymerase chain reaction (PCR) (Hamada and Kakunaga 1982; Litt and Luty 1989; Weber and May 1989; Love et al. 1990). They are locus-specific and widely dispersed, occurring every 30 kb in the euchromatin region in humans (Stallings et al. 1991).

Mouse clones have a higher frequency of $(GT)_n$ sites in cosmid screening—78% compared to 63% in humans. It is estimated that in mouse, they occur every 18 kb. Microsatellites including $(GA)_n$, $(TC)_n$, and mononucleotide repeats (Aitman et al. 1991; Cornall et al. 1991a; Hearne et al. 1991) represent some of the most polymorphic markers to date. Other polymorphism detection methods include those for single-base changes in DNA, which are rapidly becoming of more use and will ultimately be the fastest means for detecting gene mutations, as well as providing a polymorphism resource (Burmeister et al. 1991; Kovar et al. 1991).

In mouse genetics, interspecific crosses between *Mus spretus* and *Mus musculus* have greatly facilitated linkage analyses. Their genomes have diverged over 3 million years so that most probes and microsatellites are polymorphic. They are thus more powerful in mapping studies than intraspecific crosses. This shows another advantage of using mouse crosses to hunt for complex disease genes (Avner et al. 1988).

High-resolution linkage mapping

The applicability of multiple-locus ASP analysis and exclusion mapping for complex traits has been studied in IDDM in humans by Hyer et al. (1991). A detailed map of polymorphic markers was used. A region (11q) known to have homology with a NOD chromosome 9 gene for which there is preliminary evidence of linkage (Prochazka et al. 1987) was analyzed. Lathrop and Lalouel (1992) had already shown that the ASP method and standard linkage analysis gave the same results for a variety of models provided a large number of multiplex families were studied.

Significantly, affection status of nondiabetic children was considered unknown, and heterogeneity according to Smith (1963) was incorporated. A LOD score criterion of 3 was used, but Lathrop and Lalouel correctly stated that this did not mean a significance level (α error) of 0.05 (see below under Multiple testing).

For given ibd_i, the maximum expected location score (ELOC) was calculated for a theoretical family with two affected sibs and flanking markers. The sample size needed for a location score of 3 is 3/ELOC. Similarly, if ELOC for an unlinked family is calculated, given ibd_i, when the expected ibd_i values are Mendelian, the required number of families is –2/ELOC to exclude linkage.

The final results are given in simple graphic form to show the number of sib pairs needed to confirm or exclude linkage in any complex disease. This is dependent on p_2 (number of sib pairs sharing two alleles identical by descent) and flanking markers. For example, sample calculations show that with 200 ASPs and flanking markers at 20%, all genetic models can be excluded with $p_2 > 0.3$.

The LOD scores were maximized as a function of θ under 250 different genetic models with different degrees of penetrance, gene frequencies and genetic heterogeneity. More than 90% of the 11q region could be excluded with a location score criterion of less than –2 for all models with $p_2 > 0.5$.

Affected pedigree method and identity by state

Lange (1986) developed identity by state (ibs) methods as an extension of ibd. The principle involves analyzing the distributional properties of a summated statistic, Z, found by determining the number of marker alleles concordant by state for relative pairs. This is independent of ibd calculations. The advantages are that, like ibd, no assumption is made on the inheritance pattern. But unlike ibd, ibs techniques can be used where polymorphism of a marker is too low to obtain informative results. Additionally, parental typing is not necessary, which is useful, for example, in late-onset disease.

Lange developed a test statistic to look for sib pair concordance compared to expected patterns and then applied it to different pedigree members (Weeks and Lange 1988). Obviously, ibs methods will be more applicable to distant relative pairs than to sib pairs. Bishop and Williamson (1990) extended analyses further and looked at the power of the test with respect to marker polymorphism, θ, and mode of inheritance of the trait. Conclusions somewhat similar to those of Risch (1990a,b,c) were reached.

Controversy has surrounded complex disease gene mapping in general, and nowhere is it more pronounced than with Alzheimer's disease. There are at least two linkages (chromosomes 19 and 21) that have been found and again refuted. Two recent papers clarify some of the data. Farrer et al. (1991) used complex segregation analysis in this condition to show a major gene effect which is dominant or codominant. A second major gene is implicated by high heterozygote transmission probabilities. Furthermore, a polygenic background is evident.

Pericak-Vance et al. (1991) studied families using the affected pedigree method (APM) disregarding unaffecteds and detected linkage to chromosome 19. This linkage was not apparent with standard maximum likelihood methods. When the latter analysis was redone disregarding unaffecteds, the linkage to chromosome 19 was seen. Haines (1991) concludes that the studies show that chromosome 21 is linked to the early-onset group (<65) and is likely to be autosomal dominant with at least one age-dependent highly penetrant locus, whereas both chromosomes 19 and 21 may be implicated in the late-onset group (>65). The mode of inheritance of these latter loci is unclear.

Quantitative trait strategies

Wilson et al. (1991), in analyzing hypertension by ASP methods, looked at quantitative traits related to the condition. They define a genometric approach (as opposed to the standard phenometric approach), which is more useful when an underlying locus is responsible for a small portion of the phenotypic variation. The variation is then related to different genotypes at that locus. White and Lalouel (1987) had advocated a similar procedure in stating that biological correlates of the phenotypic trait should be separately analyzed. Wilson et al. show how to analyze quantitative data simply. A marker-phenotype matrix is constructed with marker loci as rows and trait phenotypes as columns. The entries are p values of significance of association for marker and trait. Then, both multilocus models (many significant p values for a phenotype column) and pleiotropy (many significant p values for a marker row) can be assessed.

Lander has developed and incorporated into MAPMAKER many aspects that are important for detection of quantitative trait loci (QTL). The algorithm reduces computer time needed to calculate likelihoods, and several QTL have been mapped in interspecific crosses of tomato (Patterson et al. 1991).

MAPMAKER has been used to find genes in complex partial seizures in mice for a seizure-prone strain (Rise et al. 1991). Multilocus mapping was used for mapping of markers in a backcross to an epilepsy-resistant strain. Seizure frequency was shown to have a continuous variation in these animals. Association between this phenotype and markers was tested in two ways: (1) a t test for mean seizure frequency between the two genotypes, regardless of genotype frequencies (quantitative analysis), and (2) a discrete analysis where mice with a seizure frequency above a threshold level were classed as seizure prone. χ^2 analysis for independence was performed between markers and phenotype status. Interval mapping was used to position a chromosome 9 susceptibility gene, and QTL analysis showed that 50% of the total phenotypic variance could be explained by the chromosome 9 locus and a putative chromosome 2 locus.

Goldgar (1990) has attempted to find through multipoint analysis the amount of genetic material shared between sibs ($=R$). $D = 1-R$, or the proportion not shared is found by the sum of the differences of map distances from the beginning of the chromosome for each crossover point. The variance and mean are easily found for R. The genetic variance of the quantitative trait is then partitioned into effects of loci on specific chromosomal regions. Three types of variations should be detected: variation due to a single major gene, variation due to several linked loci, and variation due to unlinked loci. Although many parameters were simplified (e.g., only two alleles at each locus) or ignored (interference, epistasis), this is a breakthrough in detecting QTL in humans.

Multiple testing

One very important question in linkage analysis of complex diseases concerns the number of models tested and parameters estimated. This may lead to misclassification and inflation of the LOD score. Multilocus linkage analysis may also add to the complexity. What is the final significance level of the analyses, i.e., the false-positive rate? This is related to the theoretical prior probabilities of linkage which, for a condition with an unknown mode of inheritance, is not easily derived. Simulation studies have been presented (Green 1990; Ott 1990) to give an idea about the significance level.

Ott (1985) had already shown that in dealing with fewer than 100 markers, the increased prior probability of true linkage in multiple typing is nullified by the increased chance of false positives. No adjustment to LOD score criteria is then necessary.

Risch (1991a) reviewed the situation and showed that, contrary to expectation, multiple marker testing decreases the posterior false-positive rate (ϕ) among significant tests. However, this fact also depends on the parameters for the marker used in question (α_g and β_g) and is true for both monogenic and polygenic traits. In complex diseases, heterogeneity, epistasis, phenocopies, etc., will all reduce $1-\beta$ (power) and inflate apparent θ and ϕ. Therefore, increase in the LOD score criteria is warranted in complex diseases but not because multiple markers are tested (this almost never increases ϕ).

A calculation for multiple model testing (if the true model is included and provided α is small compared to the number of tests, t) shows that φ is directly proportional to t. Therefore, a correction has to be made $\alpha' = \alpha/t$ and the significance level has to be altered, LOD [3 + $\log_{10}(t)$].

Limiting factors in complex studies

It is clear from the examples and discussion so far that mapping of complex disease genes is feasible in general provided (1) there are many, evenly spread, highly polymorphic markers; (2) there are a large number of human pedigrees or experimentally crossed animals; and (3) there is insubstantial genetic heterogeneity. If the latter is untrue, power to detect linkage by standard analysis is decreased, although large numbers of subjects will help to alleviate this problem.

The first point will not be a serious limitation in the foreseeable future. We have no control over the third factor, but new techniques of statistical analysis are already having a major impact (see above under Genetic heterogeneity). Furthermore, studying an inbred animal model will reduce the amount of interlocus heterogeneity one must analyze. The second problem is the most difficult in humans and will impede progress.

GENETIC ANALYSIS OF IDDM
Epidemiology

IDDM has a prevalence in the human population of 0.3%. Genetically identical twins are 36% concordant, and the average risk to siblings for probands is 6%. The MHC locus is a gene complex on chromosome 6 heavily implicated in disease etiology, and MHC-identical siblings are about 12% concordant (Leigh-Field 1988; Todd 1990b).

The environment therefore has a major role to play in IDDM, but MHC only accounts for a proportion of the genetic contribution. Knowing λ_R from population studies, one can calculate p_0 and compare this with values obtained from sib pair analysis (Risch 1987). For IDDM and MHC, $\lambda_s = 15$, $p_0 = 0.25/15 = 0.0167$, which is far lower than the observed value of 0.073 (45 of 612 pairs). The value of 0.073 gives estimated $\lambda_e = 0.25/0.073 = 3.42$. Therefore, 15/3.42 = 4.38 represents an increase in risk to sibs due to an undefined locus (loci), provided these are considered to be acting multiplicatively. This does not prove the existence of other genetic loci but does strongly suggest their presence.

Significant heterogeneity for θ has always been apparent in linkage studies for IDDM and HLA. DR3/4 individuals have shown a closer linkage than other diseased haplotypes. This suggests that in non-DR3/4 individuals, genetic influences outside the HLA which contribute cause an apparent increase in recombination frequency between HLA and disease.

The most important marker outside HLA implicated to date has been the 5' region of the insulin gene. Many studies have confirmed a population association between a specific type of insulin 5' allele (class 1) and IDDM. However, this has never shown as linkage in standard sib pair analysis. This confounding result has been explained by saying that there is a large sporadic component to insulin gene predisposition.

Leigh-Field (1991) circumvents the question of controls in association studies by devising a method known as association analysis of family data (AFBAC or affected family based controls). The system uses an intrafamily control method. Simply, parental haplotypes are divided into a D0 group if the haplotype does not occur in any diseased member, D1 if it does, and D2 if the haplotype occurs in more than two affected members (extended method). Comparison of the different groups in terms of haplotype frequencies will give an accurate estimation of "haplotype usage" in diabetes. The haplotype relative risk method (Falk and Rubinstein 1987) is a similar analytic method for simplex families.

Using AFBAC analyses, Thompson et al. (1989) showed that for sib pair haplotype sharing of class-1 alleles, there is a significant distortion, with 83% in the D1 category and 69% in the D0 ($p < 0.01$). Finally, it was proven that the insulin-IGF2 region has a definite role to play in diabetes etiology.

G.M. Lathrop (pers. comm.) has recently suggested an explanation for this paradox of population association without evidence for linkage: HLA DR4-positive diabetic sibs share the father's alleles for the insulin 5' VNTR more often than the mother's. Thus, this preferential transmission of IDDM-associated alleles (possibly due to maternal imprinting on 11p, a well-established phenomenon associated with the IGF2 locus) to HLA DR4-positive diabetic offspring will increase allele sharing from this subgroup, giving a weak population association but not enough to give a significantly distorted sib pair sharing distribution overall. This interesting HLA/INS locus interaction, only apparent in DR4 diabetics, and the imprinting on 11p (DeChiara et al. 1991) suggest another mechanism in complex diseases for the distortion of Mendelian segregation.

NOD mouse genetics

We have focused on NOD mice because they are the animal of choice for genetic analysis. The BB (bio-breeding) rat, a type-1 diabetic model, is not discussed (see Jackson et al. 1984; Markholst et al. 1991). Friedman et al. (1991) discuss type-2 diabetic models.

IDDM is an autoimmune disease in which the body's immune system selectively destroys the insulin-producing β cells of the pancreas. This slow, insidious process leads eventually to frank diabetes. Diabetes in the NOD mouse has many similarities to the human disease. The development of inflammation within pancreatic islets (insulitis) is thought to match the human pattern, although we have to take into consideration the different life spans for humans and mice. Female NOD mice develop a cumulative incidence of disease of 0.9, and males, 0.5 at 330 days. This sex discordance in NOD mice is a major difference from that in humans. NOD is inbred and can readily be outcrossed for mapping studies.

Hattori et al. (1986) analyzed the MHC of NOD mice. The haplotype is $H-2K^d$ $H-2D^b$ for class-1 antigens and $I-A^{nod}$ $I-E_\alpha^o I-E_\beta^+$ for class 2. NOD has a unique $I-A_\beta$ chain (Acha-Orbea and McDevitt 1987). Hattori's outcross of NOD was with C3H mice ($H-2^k$), and he backcrossed onto NOD; 15.6% of the mice in the backcross became diabetic of a total of 135 and all diabetics were homozygous for MHC. These authors suggested the existence of at least one and perhaps two genes outside the MHC controlling diabetes.

Prochazka et al. (1987) confirmed that NOD MHC homozygosity is essential for diabetes. They also went on to show from a NOD backcross after an outcross to NON that perhaps a single recessive gene controlled the development of insulitis. "Linkage" to a putative locus outside the MHC in their 19 diabetic mice (out of 200 progeny) in the backcross was based on 3 of 19 animals being heterozygous for *Alp-1* on chromosome 9. This locus was associated with diabetes but, importantly, not insulitis.

In outcrossing NOD to IL1 (which incidentally is the only strain having the same MHC class-1 and -2 genes as NOD) and backcrossing to NOD, Hattori et al. (1990) estimated insulitis incidence to be 40%. All backcross animals have the same diabetogenic homozygous MHC; this is an important scheme both in trying to increase the incidence of diabetes and in excluding the segregation of MHC with diabetes or insulitis. In backcrosses with C3H and NON, females had an incidence of insulitis of 50%, regardless of MHC heterozygosity or homozygosity. These three crosses show that a major gene outside the MHC controls insulitis. Had MHC controlled insulitis recessively (full penetrance), then the trait would be present at 100% in the IL1 cross. No linkage to chromosome 9 for insulitis was seen.

A highly perceptive analysis of the immunogenetics of type-1 diabetes was performed by Wicker et al. (1987). Whereas, previously, modeling of the interacting diabetes susceptibility genes had been assumed to be multiplicative with full penetrance, this group was far more open-minded about the possibilities. The other strain in their studies was C57BL/10 (B10), and F_1, F_2, and F_1 x NOD first, second, and third backcross generations were constructed. NOD homozygosity was shown to be almost a prerequisite for diabetes as well as influencing the severity of insulitis. There were estimated to be at least two genes not linked to the MHC influencing diabetes that gave the best fit model for the observed data. One influenced the initiation of insulitis and was incompletely dominant. The other gene was purported to control T-suppressor cell function and to lead from severe insulitis to diabetes; hence, it would be linked to diabetes but not insulitis.

Recently, congenic experiments have shown that NOD mice homozygous for H-2 of NON do not develop insulitis. (In these experiments, inbred mice are made homozygous for one locus of a separate strain; all other background genes are supposed to be of the primary strain. In the breeding methods to construct these mice, however, random loci may also be inadvertently made congenic for the donor strain [Prochazka et al. 1989].)

To define the role of the MHC further, Wicker et al. (1991a,b) have produced NOD mice congenic for $H\text{-}2^b$(NOD $H\text{-}2^b$ N6). There is no diabetes and almost no insulitis (1/22 animals shows mild inflammation in which less than 25% of the islets contain infiltrating cells). Insulitis is very rare in the total absence of NOD MHC, but the latter is not the only major factor, as B10.$H\text{-}2^{g7}$ (B10 made congenic for NOD MHC) has no insulitis.

Because MHC heterozygotes have a lower incidence of diabetes (3% in a [NOD x B10] x NOD cross), it is assumed that MHC homozygotes have a higher penetrance. The insulitis present in these heterozygotes, after many backcross matings with NOD, still shows extreme variability in expression (Wicker et al. 1989). One assumes that this can only be due to environmental differences. This has been corroborated in another cross (NOD x SWR)F_1 x NOD by Livingstone et al. (1991), who found 17 diabetics in a backcross of 213; 3 of 17 were heterozygous for NOD MHC and the rest were homozygous. Insulitis is not dependent on MHC heterozygosity. Recent work by Prochazka et al. (1989) has confirmed the permissive role that MHC plays in insulitis and the essential role it plays in diabetes.

Microsatellite linkage map

We started our analysis first by confirming the abundance of microsatellites in mice and second by showing them to be highly variable even between inbred strains (Love et al. 1990). There are five main sources for microsatellites.

1. Computer data bases: EMBL and GENBANK DNA data bases were searched for >$(CA)_{10}$, >$(GT)_{10}$, >$(GA)_{10}$, >$(CT)_{10}$, and mononucleotide repeats (Love et al. 1990; Aitman et al. 1991; Hearne et al. 1991). More complex repeats were detected using the MAKEPAT program and NIPL (Staden, Cambridge).

2. Cloned DNA: A directed approach has been utilized to screen and isolate microsatellites from previously cloned genomic DNA in phage and cosmid vectors.
3. Genomic libraries: A random approach of screening phage libraries constructed from genomic DNA digested with frequently cutting enzymes was used (Cornall et al. 1991a).
4. Flow sorted libraries: In collaboration with Dr. J. Friedman, we have screened a phage library constructed from flow-sorted 4/6 Robertsonian chromosomes (unpubl.).
5. Interspersed repeat sequence amplification: Using LINE and SINE consensus site repeat primers, we have amplified and cloned chromosome-specific microsatellites from somatic cell hybrids (unpubl.).

To date, more than 230 microsatellites have been studied in our laboratory. Of those that gave a clear product (207), 76% (158/207) have at least two alleles between any two strains, and 70% (144/207) between C57B6/J (B6/J) and *Mus spretus* (SPE); 26% (54/207) show variation between B10.H-2^{g7} (C57BL/10SnJ made congenic for *H-2* region of NOD) and NOD. The latter were the progenitor strains for our backcross.

Mapping markers

Once a microsatellite is isolated and PCR conditions are optimized, it must be mapped (if this has not already been done). The Jackson map (Jackson Laboratory, Bar Harbor, Maine) was used for microsatellites from known locations.

Strain distribution patterns (SDP) across recombinant inbred strains (RIS) constitute a powerful method, especially between markers a short distance apart. If the marker is variant between the progenitor strains, then mapping of a marker is possible across the RIS by comparing with other markers of known patterns. The accuracy over shorter distances lies in the fact that 20 generations of brother-sister matings are needed to produce each of the 30 strains; more meioses are studied than in a single cross.

Intra- and interspecific crosses were used to map markers as well. These were larger than the numbers in the RIS and proved to be useful adjuncts. The power of the interspecific approach was that most markers (70%; see above) could be mapped in (SPE x B6/J) XB6/J.

Diabetic cross

The major intraspecific cross used consisted of two reciprocal backcrosses (BC1). (B10.H-2^{g7} x NOD)F$_1$ x NOD generated 59/430 (13.7%) female diabetic mice, and NOD x (B10.H-2^{g7} x NOD)F$_1$ generated 42/282

(14.9%). Livers were used for DNA extraction, and pancreas histology was performed to grade insulitis in a proportion of the controls. Each marker variant in the progenitor strains was run across the backcross DNA. These included 8 RFLP variants and 53 locus-specific microsatellites. There was at least one marker per chromosome.

Statistical analysis

For each marker locus, differences in the NOD/NOD (NN) homozygote and B10/NOD (NB) heterozygote frequencies in diabetic and control animals were evaluated by a χ^2 test of independence. When significant evidence of nonrandom segregation was found ($\chi^2 > 13.8$), placement of the IDDM susceptibility gene was initially estimated under a model that assumed a single gene with unknown penetrance. Genetic distances were calculated using the Kosambi mapping function, and gene order was confirmed by the GMS program of the LINKAGE package (Lathrop et al. 1988).

Location estimates were obtained by a modification of the location score approach (Lathrop et al. 1984). The location score method in its original form first uses multipoint linkage analysis to fix recombination estimates between markers. Then the likelihood is made a function of the subjective placement of a disease locus, given penetrance. A support test is constructed to give a location score (as twice the natural logarithm of the odds ratio).

Another method is to give maximum likelihood estimates of recombination and location together. In our case, penetrance was made a function of age by a logistic function, and the estimates for the parameters of this model were made at each location. The two methods gave the same answers for relative positions of disease loci.

Key mapping data

Idd-1 refers to the MHC as a diabetes susceptibility locus, and *Idd-2* refers to the putative chromosome 9 region. We discuss two new genes, *Idd-3* and *Idd-4*, found from our analysis. *Idd-3* (chromosome 3) is localized to an area spanned by *Il-2*, *Tshb*, and *D3Nds1* ($\chi^2 = 30.4$). Linkage to *Idd-4* ($\chi^2 = 15.8$ for young animals at *D11Nds1*, chromosome 11) is affected by age of onset more than *Idd-3*.

There are 9 of 97 diabetic animals heterozygous for *Il-2*, *D3Nds1*, and *Tshb*. Similarly, 4 of 32 early-onset diabetic animals are heterozygous across *Acrb*, *D11Nds1*, and *Mpo* (chromosome 11). This implies that the diabetic trait does not have a simple recessive mode of inheritance at these loci. Homozygosity for NOD at *D3Nds1* has a strong association with severity of insulitis ($\chi^2 = 11.7$; $p<0.001$). There is no such correlation with any chromosome 11 marker. *Idd-4* may be the gene that Wicker proposed would cause progression from insulitis to diabetes.

No animal is heterozygous and still diabetic across both chromosomal areas for the disease susceptibility genes. This implies that the genes have some form of interaction, and preliminary analyses point to a mixed additive and multiplicative model. We estimate from our data that 70% of the genome is covered in terms of detecting genes with effects equivalent to *Idd-4*.

More recently, we have localized a third non-MHC gene, *Idd-5*, to mouse chromosome 1, which influences both insulitis and diabetes (Cornall et al. 1991b). These appear to be multiple loci acting in this backcross reflecting the complexity of the inheritance of diabetes even within inbred mouse strains.

SUMMARY AND THE STUDY OF HUMAN IDDM

Recent advances in statistical and molecular techniques have made possible the systematic study of the genetics of complex disease. We and other investigators (Neumann and Collins 1990; Rise et al. 1991) have shown that by having a dense map of variant markers and a large mouse cross, complex disease loci can be mapped. The construction of congenic strains will allow both fine mapping and assessment of the function of disease susceptibility genes.

Furthermore, the extensive mouse/human homology maps make this mode of study an attractive starting point for tackling the even more difficult problems that will inevitably arise in the study of complex disease in outbred human populations.

Our approach to human IDDM is not a total human genome search; rather, we concentrate on regions homologous to disease susceptibility genes found in the mouse and possible candidate gene loci involved in susceptibility to autoimmunity.

1. We study 200 multiplex nuclear families (Hyer et al. 1991) with two or more affected sibs. A repository of Epstein-Barr-virus-transformed cells provides a long-term source of DNA from these families.
2. We divide diabetics into age-of-onset groups of less than 17 and greater than 17 to do separate genetic analyses.
3. The markers employed in linkage analysis are microsatellites and variable number tandem repeats with PIC >0.7. Eventually, we hope to have a map density of these markers every 20 cM.
4. We ignore unaffecteds in the analysis to increase the chance of detecting linkage, although they are typed to detect transmission distortion of alleles.

5. We study other relative pairs besides sibs. Distant relatives offer greater power to detect linkage in the presence of the complicated multilocus effects (Risch 1990b).
6. We also perform population association studies with candidate genes and marker loci linked to disease using patients and a large panel of unrelated, ethnically and sex-matched controls.

Acknowledgments

We thank the Wellcome Trust, the Medical Research Council, as part of the U.K. Human Genome Mapping Project, the Juvenile Diabetes Foundation, and the British Diabetes Association. We are also grateful to G.M. Lathrop, P. Neumann, M. Boehnke, and S.C. Bain for many helpful suggestions.

References

Acha-Orbea, H. and H.O McDevitt. 1987. The first external domain of NOD class 2 IAβ chain is unique. *Proc. Natl. Acad. Sci.* **82:** 2435.

Aitman, T.J., C.M. Hearne, M.A. McAleer, and J.A. Todd. 1991. Mononucleotide repeat sequences are an abundant source of length variants in mouse genomic DNA. *Mamm. Genome* **1:** 206.

Avner, P., L. Amar, L. Dandalo, and J.L. Guenet. 1988. Genetic analysis of mouse using interspecific crosses. *Trends Genet.* **4:** 18.

Bishop, D.T. and J.A. Williamson. 1990. The power of identity by state methods for linkage analysis. *Am. J. Hum. Genet.* **46:** 254.

Boehnke, M., K.H. Omoto, and J.M. Arduino. 1990. Selecting pedigrees for linkage analysis of a quantitative trait: The expected number of informative meioses. *Am. J. Hum. Genet.* **46:** 581.

Bonney, G. 1986. Regressive logistic models for familial disease and other binary traits. *Biometrics* **42:** 611.

Botstein, D., R.L. White, M. Skolnick, and R.W. Davis. 1980. Construction of a genetic linkage map in man using RFLP's. *Am. J. Hum. Genet.* **32:** 314.

Burmeister, M., G. diSibio, D.R. Cox, and R.M. Myers. 1991. Identification of polymorphisms by genomic denaturating gradient gel electrophoresis: Application to the proximal region of chromosome 21. *Nucleic Acids Res.* **19:** 1475.

Clerget-Darpoux, F. and C. Bonaiti-Pellie. 1980. Epistasis effect: An alternative to the hypothesis of linkage disequilibrium in HLA associated disease. *Ann. Hum. Genet.* **44:** 195.

Cornall, R.J., T.J. Aitman, C.M. Hearne, and J.A. Todd. 1991a. The generation of a library of PCR-analyzed microsatellite variants for genetic mapping in the mouse genome. *Genomics* **10:** 874.

Cornall, R.J., J.-B. Prins, J.A. Todd, A. Pressey, N.H. DeLasato, L.S. Wicker, and

L.B. Peterson. 1991b. Type 1 diabetes in mice is linked to the interleukin-1 receptor and *Lsh/Ity/Bcg* genes on chromosome 1. *Nature* 353: 262.

DeChiara, T.M., E.J. Robertson, and A. Efstratiadis. 1991. Parental imprinting of mouse insulin-like growth factor II gene. *Cell* 64: 849.

Edwards, A.W.F. 1972. The concept of likelihood. In *Likelihood*. Cambridge University Press, United Kingdom.

Elston, R.C. 1981. Segregation analysis. *Adv. Hum. Genet.* 10: 63.

Erickson, R.P. 1990. Mapping dysmorphic syndromes with the aid of human/mouse homology maps. *Am. J. Hum. Genet.* 46: 1013.

Falconer, D.S. 1989. Genetic constitution of a population. In *Introduction to quantitative genetics*, p. 18. Longman, New York.

Falk, C.T. and P. Rubinstein. 1987. Haplotype relative risks: An easy reliable way to construct a proper control sample for risk calculations. *Ann. Hum. Genet.* 51: 227.

Farrer, L.A., R.H. Myers, L. Connor, L.A. Cupples, and J.H. Growdon. 1991. Segregation analysis reveals evidence of a major gene for Alzheimer's disease. *Am. J. Hum. Genet.* 48: 1026.

Foy, C., V. Newton, D. Weellesley, R. Harris, and A.P. Read. 1990. Assignment of the locus for Waardenberg's syndrome type 1 to human chromosome 2q37 and possible homology to splotch mouse. *Am. J. Hum. Genet.* 46: 1017.

Friedman, J.M., R.L. Leibel, and N. Barahy. 1991. Molecular mapping of obesity genes. *Mamm. Genome* 1: 130.

Goldgar, D.E. 1990. Multipoint analysis of human quantitative genetic variation. *Am. J. Hum. Genet.* 47: 957.

Green, P. 1990. Genetic linkage and complex disease—A comment. *Genet. Epidemiol.* 7: 25.

Haines, J.L. 1991. Genetics of Alzheimer's disease—A teasing problem. *Am. J. Hum. Genet.* 48: 1021.

Hamada, H. and T. Kakunaga, 1982. Potential z DNA forming sequences are highly dispersed in the human genome. *Nature* 298: 396.

Hattori, M., M. Fukuda, T. Ichikawa, H.J. Baumgartl, H. Katoh, and S. Makino. 1990. A single non-MHC diabetogenic gene determines the development of insulitis in the presence of an MHC-linked diabetogenic gene in NOD mice. *J. Autoimmun.* 3: 1.

Hattori, M., J.B. Buse, R.A. Jackson, L. Glimcher, E. Dort, M. Minami, S. Makino, K. Moriwaki, H. Kuzuya, H. Imura, W.M. Strauss, J.G. Seidman, and G.S. Eisenbarth. 1986. The NOD mouse: Recessive diabetogenic gene in the major histocompatibility complex. *Science* 231: 733.

Hearne, C.M., S. Ghosh, and J.A. Todd. 1992. Microsatellites for linkage analysis of complex human disease. *Trends Genet.* (in press).

Hearne, C.M., M. McAleer, J.M. Love, T.J. Aitman, R.J. Cornall, S. Ghosh, A.M. Knight, J.B. Prins, and J.A. Todd. 1991. Additional microsatellite markers for mouse genome mapping. *Mammal. Genome* 1: 273.

Hilbert, P., K. Lindpaintner, J.S. Beckmann, T. Serikawa, F. Soubrier, C. Dubay, P. Cartwright, B. De Gouysn, C. Julier, T. Takahasi, M. Vincent, D. Ganten, M. Georges, and G.M. Lathrop. 1991. Chromosomal mapping of two genetic loci associated with blood pressure regulation in hereditary hypertensive rats. *Nature* 353: 521.

Hodge, S.E. 1981. Some epistatic two locus models of disease. Relative risks and IBD distributions in affected sib pairs. *Am. J. Hum. Genet.* 37: 381.

Hyer, R.N., C. Julier, J.D. Buckley, M. Trucco, J. Rotter, R. Spielman, A. Barnett, S. Bain, C. Boitard, I. Deschamps, J.A. Todd, J.I. Bell, and G.M. Lathrop. 1991. High resolution linkage mapping for susceptibility genes in human polygenic disease. IDDM and chromosome 11q. *Am. J. Hum. Genet.* **48:** 243.

Jackson, R.A., J.B. Buse, R. Rifai, D. Pelletier, E.L. Milford, C.B. Carpenter, G.S. Eisenbarth, and R.M. Williams. 1984. Two genes required for diabetes in BB rats—Evidence from cyclical intercrosses and backcrosses. *J. Exp. Med.* **159:** 1629.

Jacob, H.J., K. Lindpaintner, S.E. Lincoln, K. Kusumi, R.K. Bunker, V.-P. Mao, D. Ganten, V.J. Dzau, and E.S. Lander. 1991. Genetic mapping of a gene causing hypertension in the stroke-prone spontaneously hypertensive rat. *Cell* **67:** 213.

James, J.W. 1971. Frequency in relatives for an all or none trait. *Ann. Hum. Genet.* **35:** 47.

Janssen, L.A.J., L.A. Sandkuyl, E.C. Merkens, J.A. Maat-Kievit, J.R. Sampson, P. Fleury, R.C.M. Hennekam, G.C. Grosveld, D. Lindlout, and D.J.J. Halley. 1990. Genetic heterogeneity in tuberous sclerosis. *Genomics* **8:** 237.

Kerem, B.S., J.M. Rommens, J.A. Buchanan, D. Markiewicz, K.T. Cox, A. Chakravorti, M. Buchwald, and L.C. Tsui. 1989. Identification of the cystic fibrosis gene: Genetic analysis. *Science* **245:** 1073.

Kovar, H., J. Gunhild, H. Awer, T. Skern, and D. Blaas. 1991. Two dimensional single strand conformation polymorphism analysis: A useful tool for detection of mutations in long DNA fragments. *Nucleic Acids Res.* **19:** 3507.

Lalouel, J.M., D.C. Rao, N.E. Morton, and R.C. Elston. 1983. A unified model for complex segregation analysis. *Am. J. Hum. Genet.* **35:** 816.

Lander, E.S. and D. Botstein. 1986. Strategies for studying heterogenous genetic traits in humans by using a linkage map of RFLP. *Proc. Natl. Acad. Sci.* **83:** 7353.

Lander, E.S. 1988. Complex genetic traits in humans. In *Genome analysis: A practical approach* (ed. K. Davies), p. 171. IRL Press, Oxford.

Lander, E.S., P. Green, J. Abrahamson, A. Barlow, M.J. Daly, S. Lincoln, and L. Newberg. 1987. MAPMAKER—An interactive computer package for constructing primary genetic linkage maps of experimental and natural populations. *Genomics* **1:** 174.

Lange, K. 1986. A test statistic for affected sib pair set method. *Ann. Hum. Genet.* **50:** 283.

Lathrop, G.M. and J.M. Lalouel. 1992. Statistical methods in linkage analysis. In *Handbook of statistics*, vol. 8 (ed. D.C. Rao and R. Chakraborty). Elsevier, Amsterdam. (In press.)

Lathrop, G.M., J.M. Lalouel, C. Julier, and J. Ott. 1984. Strategies for multilocus linkage analysis in humans. *Proc. Natl. Acad. Sci.* **81:** 3443.

Lathrop, G.M., Y. Nakamura, P. Cartwright, P. O'Connell, M. Leppert, C. Jones, H. Tateishi, T. Bragg, J.M. Lalouel, and R. White. 1988. A primary genetic map of markers for human chromosome 10. *Genomics* **2:** 157.

Leigh-Field, L. 1988. Insulin dependent diabetes mellitus: A model for the study of multifactorial disorders. *Am. J. Hum. Genet.* **43:** 793.

———. 1991. Non-HLA region genes in IDDM. In *Balliere's clinical endocrinology and metabolism* (ed L.C Harrison and B.D. Tait), p.413. Balliere Tindall, London.

Litt, M. and J.A. Luty. 1989. A hypervariable microsatellite revealed by in-vitro amplification of a dinucleotide repeat within the cardiac muscle actin gene. *Am. J. Hum. Genet.* **44:** 397.

Livingstone, A., C.T. Edwards, J.A. Shizuru, and C.G Fathman. 1991. Genetic analysis of diabetes in NOD mouse: MHC and T-cell receptor β gene expression. *J. Immunol.* **146:** 529.

Love, J.M., A.M. Knight, M. McAleer, and J.A. Todd. 1990. Towards construction of a high resolution map of the mouse genome using PCR analysed microsatellites. *Nucleic Acids Res.* **18:** 4123.

Markholst, H., S. Eastman, D. Wilson, B.E. Andreason, and A. Lernmark. 1991. Diabetes segregates as a single locus in crosses between BB rats diabetes prone or resistant to diabetes. *J. Exp. Med.* **174:** 297.

Morton, N.E. 1990. Genetic linkage and complex disease—A comment. *Genet. Epidemiol.* **7:** 33.

Nadeau, J.H. 1989. Maps of linkage and synteny homologies between mouse and man. *Trends Genet.* **5:** 82.

Nakamura, Y., M. Leppert, P. O'Connell, R. Wolff, T. Holm, M. Culver, C. Martin, E. Fujimoto, M. Hoff, E. Kumlief, and R. White. 1987. Variable number of tandem repeats (VNTR) markers for human gene mapping. *Science* **235:** 1616.

Neumann, P.E. and R.L. Collins. 1991. Genetic dissection of susceptibilty to audiogenic seizures in inbred mice. *Proc. Natl. Acad. Sci.* **88:** 5408.

Ott, J. 1985. Penetrance. In *Analysis of human genetic linkage*, p. 132. John Hopkins University Press, Baltimore.

———. 1990. Genetic linkage and complex disease—A comment. *Genet. Epidemiol.* **7:** 35.

Patterson, A.H., S. Damon, J.D. Hewitt, D. Zamir, H.D. Rabinowitch, S.E. Lincoln, E.S. Lander, and S. Tanksley. 1991. Mendelian factors underlying quantitative traits in tomato; comparison across species, generations and environments. *Genetics* **127:** 181.

Pericak-Vance, M.A., J.L. Bebout, P.C. Gaskell, Jr., L.H. Yamaoka, W.Y. Hung, M.J. Alberts, A.P. Walker, R.J. Bartlett, C.A. Haynes, K.A. Welsh, N.L. Earl, A. Heyman, C.M Clark, and A.D. Rosen. 1991. Linkage studies in familial Alzheimer's disease: Evidence for chromosome 19 linkage. *Am. J. Hum. Genet.* **48:** 1034.

Prochazka, M., E.H. Leiter, D.V. Serreze, and D.L. Coleman. 1987. Three recessive loci required for insulin-dependent diabetes in nonobese diabetic mice. *Science* **237:** 286.

Prochazka, M., D.V. Serreze, S.M. Worthen, and E.H. Leiter. 1989. Genetic control of diabetogenesis in NOD/Lt mice. Development and analysis of congenic stocks. *Diabetes* **38:** 1446.

Risch, N. 1984. Segregation analysis incorporating linkage markers 1: Single locus models with an application to type 1 diabetes. *Am. J. Hum. Genet.* **36:** 363.

———. 1987. Assessing the role of HLA linked and unlinked determinants of disease. *Am. J. Hum. Genet.* **40:** 1.

———. 1990a. Linkage strategies for genetically complex traits: Multilocus models. *Am. J. Hum. Genet.* **46:** 222.

———. 1990b. The power of affected relative pairs. *Am. J. Hum. Genet.* **46:** 229.

———. 1990c. The effect of marker polymorphism on the analysis of affected relative pairs. *Am. J. Hum. Genet.* **46**: 242.

———. 1991a. A note on multiple testing procedures in linkage analysis. *Am. J. Hum. Genet.* **48**: 1058.

———. 1991b. Developments in gene mapping with linkage methods. *Curr. Opin. Genet. Dev.* **1**: 93.

Rise, M.L., W.N. Frankel, J.M. Coffin, and T.N. Seyfried. 1991. Genes for epilepsy mapped in mouse. *Science* **253**: 669.

Schork, N. 1991. Efficient computation of patterned covariance matrix mixed models in quantitative segregation analysis. *Genet. Epidemiol.* **8**: 29.

Smith, C.A.B. 1963. Testing for heterogeneity of recombination values in human genetics. *Ann. Hum. Genet.* **27**: 175.

Stallings, R.L., A.F. Ford, D. Nelson, D.C. Torney, C.E. Hildebrand, and R.K. Moyzis. 1991. Evolution and distribution of (GT)n repeat sequences in mammalian genomes. *Genomics* **10**: 807

Tait B.D. and L.C. Harrison. 1991. Overview MHC and IDDM. In *Balliere's clinical endocrinology and metabolism* (ed L.C Harrison and B.D. Tait), p. 211. Balliere Tindall, London.

Thompson, G., W.P. Robinson, M.K. Kuhner, S. Joe, and W. Klitz. 1989. HLA and insulin gene associations with IDDM. *Genet. Epidemiol.* **6**: 155.

Thompson, G., W.P. Robinson, M.K. Kuhner, S. Joe, M.J. McDonald, J.L. Gottschall, J. Barbosa, S.S. Rich, J. Bertrams, M.P. Baur, J. Partanen, B.D. Tait, E. Schober, W.R. Mayr, J. Ludviggson, B. Lindblom, N.R. Farid, C. Thompson, and I. Deschamps. 1988. A joint study of Caucasians with IDDM. *Am. J. Hum. Genet.* **43**: 799.

Todd, J.A. 1990a. The role of MHC class 2 genes in susceptibility to IDDM. *Curr. Top. Microbiol. Immunol.* **164**: 17.

———. 1990b. Genetic control of autoimmunity in type 1 diabetes. *Immunol. Today* **11**: 122.

Todd, J.A., J.I. Bell, and H.O. McDevitt. 1987. HLA DQβ gene contributes to susceptibility and resistance to IDDM. *Nature* **329**: 599.

Todd, J.A., T.J. Aitman, R.J. Cornall, S. Ghosh, J.R.S. Hall, C.M. Hearne, A.M. Knight, J.M. Love, M.A. McAleer, J.B. Prins, N. Rodrigues, G.M. Lathrop, A. Pressey, N.H. DeLarato, L.B. Peterson, and L.S. Wicker. 1991. Genetic analysis of autoimmune type 1 diabetes mellitus in mice. *Nature* **351**: 542.

Weber, J.L. and P.E. May. 1989. Abundant class of human DNA polymorphisms which can be typed using PCR. *Am. J. Hum. Genet.* **44**: 388.

Weeks, D.E. and K. Lange. 1988. Affected pedigree member method of linkage analysis. *Am. J. Hum. Genet.* **42**: 315.

White, R. and J.M. Lalouel. 1987 Investigation of genetic linkage in human families. *Adv. Hum. Genet.* **16**: 121.

Wicker, L.S., N.H. DeLaroto, A. Pressey, and L.B. Peterson. 1991a. Genetic control of diabetes in the NOD mouse: Analysis of the NOD.H-2^b and B10.H-2^{nod} strains. In *Proceedings of P & S Biomedical Sciences Symposium: Molecular mechanisms of immunological self-recognition* (ed. H.J. Vogel and S.W. Alt). Academic Press, New York. (In press.)

———. 1991b. Speculation on the genetic control of diabetes and insulitis in the NOD mouse. In *Frontiers in diabetic research II. Lessons from animal diabetes II* (ed. E. Shafir), p. 67. Smith-Gordon, London.

Wicker, L.S., B.J. Miller, P.A. Fischer, A. Pressey, and L.B. Peterson. 1989. Genetic control of diabetes in NOD mice. Pedigree analysis of a $H2^{nod/b}$

heterozygote. *J. Immunol.* **142:** 781.

Wicker, L.S., B.J. Miller, L.Z. Coker, S.E. McNally, S. Scott, Y. Mullen, and M.C Appel. 1987. Genetic control of diabetes and insulitis in the nonobese diabetic (NOD) mouse. *J. Exp. Med.* **165:** 1639.

Wilson, A.F., R.C. Elston, L.D. Tran, and R.M. Siervogal. 1991. Use of robust sib pair method to screen for single locus multiple loci, and pleiotropic effects; application to traits related to hypertension. *Am. J. Hum. Genet.* **48:** 862.

Molecular Biology of the *W* and *Steel* Loci

Alastair D. Reith[1] and Alan Bernstein
Division of Molecular and Developmental Biology
Samuel Lunenfeld Research Institute
Mount Sinai Hospital
Toronto M5G 1X5, Canada

Mutations at the murine *Dominant white-spotting* (*W*) and *steel* (*Sl*) loci both lead to defects in hematopoiesis, melanogenesis, and gametogenesis. Recent experiments have shown that *W* is allelic with c-*kit*, a proto-oncogene that encodes a receptor tyrosine kinase (RTK), and *Sl* encodes the ligand for the Kit receptor. This chapter provides an overview of the biology, genetics, and molecular biology of these classic mammalian developmental loci.

The main topics discussed include:

❑ molecular classification of *W* mutants into those that affect the levels of gene expression (regulatory) and those that directly affect the structure of the Kit receptor (structural)

❑ molecular basis of dominant-negative *W* phenotypes

❑ biochemistry and genetics of the Kit signaling pathway, including the identification of the Kit ligand as the product of the *steel* locus

❑ *Ph-W-Rw-rs* gene cluster

❑ occurrence and phenotype of germ-line mutations in the c-*kit* gene in other mammals, including rats and humans

INTRODUCTION

Growth factor receptors with intrinsic tyrosine kinase activity constitute a large family of structurally and functionally related molecules that play

---
[1]Present Address: Ludwig Institute for Cancer Research, Courtauld Building, 91 Riding House Street, London WIP 8BT, United Kingdom.

critical roles in the regulation of cellular proliferation and the determination of cell fate (for recent reviews, see Hanks et al. 1988; Yarden and Ullrich 1988; Pawson and Bernstein 1990; Sherr 1990; Ullrich and Schlessinger 1990). Many receptor tyrosine kinases (RTKs) were first identified as a consequence of the latent oncogenic activities of this class of proteins. For example, the RTKs *neu*, *ret*, and *trk* were identified initially as chemically induced oncogenes, whereas other RTKs, including c-*erbB*, c-*ros*, c-*fms*, and c-*kit*, were identified as the cellular homologs of the viral oncogenes present in the genomes of oncogenic retroviruses.

In addition to their mitogenic roles associated with specific malignancies, RTKs have also been identified as the protein products encoded by several developmental genes that control cell fate in *Drosophila* (for review, see Rubin 1989; Pawson and Bernstein 1990).

Amino acid sequence comparisons have revealed similarities in the overall structure of RTKs and have allowed their classification into a number of subgroups (Fig. 1) (Hanks et al. 1988). Moreover, mutagenic analyses, coupled with the isolation of activated oncogenic variants of these receptors, have facilitated a detailed analysis of the molecular mechanisms by which these molecules function in signaling pathways. Ligand binding to the extracellular domain is believed to induce receptor dimerization and autophosphorylation at specific cytoplasmic tyrosine residues. Autophosphorylation acts as a switch enabling the receptor to bind and phosphorylate cytoplasmic proteins, thereby transducing an extracellular proliferative signal to the cytoplasm.

The known roles of RTKs and their ligands in the regulation of cellular proliferation have led to much interest and speculation as to their functions in mammalian development, but until recently, direct evidence was lacking. Molecular analysis of the murine W/c-*kit* and *steel* loci has provided an important experimental system in which to investigate the roles played by RTK-mediated signaling pathways in the regulation of mammalian development.

The proto-oncogene c-*kit* was first identified as the cellular homolog of the activated oncogene v-*kit* present in the genome of the acutely transforming feline retrovirus HZ4-feline sarcoma virus (FeSV). This virus induces multicentric fibrosarcomas in the domestic cat as a direct result of expression of the oncogene v-*kit*, the predicted amino acid sequence of which is homologous to the receptor class of tyrosine kinases (Besmer et al. 1986). c-*kit*, the cellular homolog of the v-*kit* oncogene, is most closely related to the platelet-derived growth factor receptor (PDGFR) class of RTKs, characterized by the presence of five immunoglobulin-like repeats in the extracellular domain and an insert that splits the cytoplasmic kinase domain into an ATP-binding region and the phosphotransferase domain (Fig. 1) (Hanks et al. 1988; Yarden et al. 1987; Qui et al. 1988). The c-*kit* locus encodes a 106-kD polypeptide

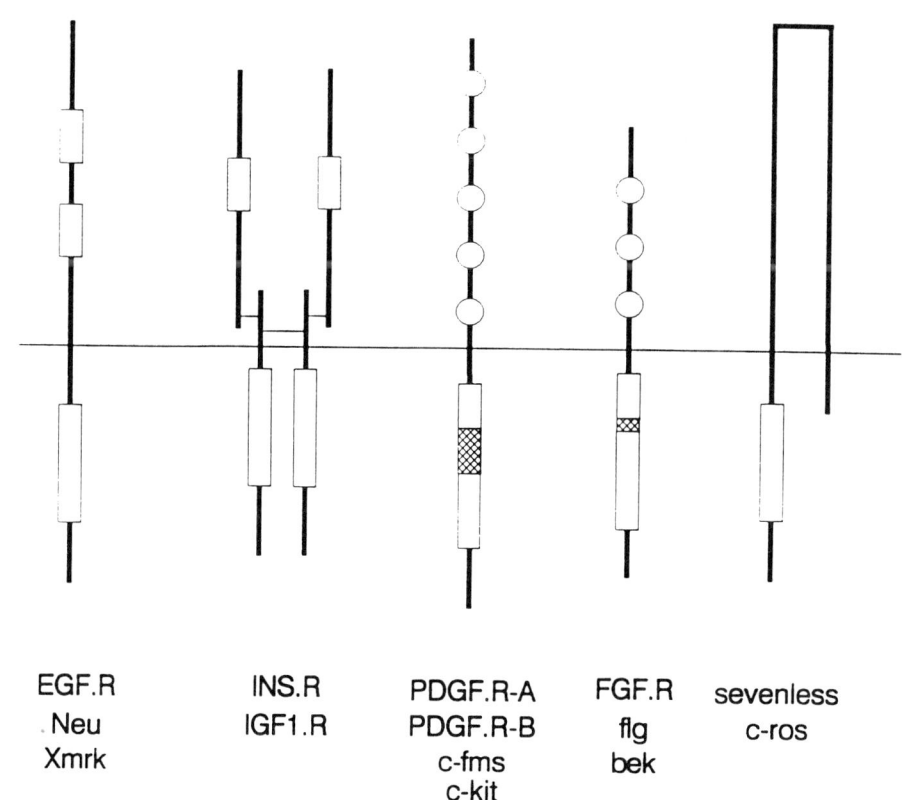

EGF.R	INS.R	PDGF.R-A	FGF.R	sevenless
Neu	IGF1.R	PDGF.R-B	flg	c-ros
Xmrk		c-fms	bek	
		c-kit		

Figure 1 The family of receptor-like protein tyrosine kinases. Structural relationships between members of the family of receptor tyrosine kinases are shown (Hanks et al. 1988; Yarden and Ullrich 1988). All possess an extracellular domain involved in recognition and binding of specific ligands, a transmembrane spanning domain, a juxtamembrane domain, a tyrosine kinase domain (open box) and a carboxy-terminal region. The platelet-derived growth factor receptor (PDGF.R) and fibroblast growth factor receptor (FGF.R) subfamilies are distinguished by the presence of a non-conserved insert region (hatched box) separating nucleotide binding and phosphotransferase elements within the kinase domain. The structure of the extracellular domain is characteristic for each class of RTKs. Cysteine-rich elements (shaded box) are found in the epidermal growth factor receptor (EGF.R) and insulin receptor (INS.R) subfamilies and five and three immunoglobulin-like domains (open circles) are found in PDGF.R and FGF.R classes, respectively. *sevenless* and *c-ros* are distinguished by the presence of a second transmembrane spanning region.

that undergoes posttranslational modifications to produce differentially glycosylated proteins of between 124 kD and 160 kD found at the cell surface and capable of autophosphorylation on tyrosine residues (Yarden et al. 1987; Majumder et al. 1988; Nocka et al. 1989).

Chromosomal mapping localized c-*kit* to human chromosome 4cen-q21 and mouse chromosome 5 as part of a syntenic group of loci including *Pgm-1*, *Alb-1*, and *Afp* (Fig. 2) (Yarden et al. 1987). Within this region of the mouse genome are a number of loci defined on the basis of the generation of mutant phenotypes, including *Dominant white spotting* (*W*), a highly mutable locus that induces dominant pleiotropic defects in cells of hematopoietic, melanogenic, and germ cell lineages (Table 1) (Russell 1979; Silvers 1979). The close linkage between *W* and *c-kit* (Chabot et al. 1988), together with the cell-autonomous nature of *W* mutations, made *c-kit* a likely candidate gene for *W*, and subsequent analyses have demonstrated that loss-of-function mutations within *c-kit* are responsible for the *W* phenotype. Molecular analysis of the repertoire of available *W* mutants has provided the first example of germ-line mutations in a mammalian RTK, facilitated an understanding of the in vivo functions of this RTK, and identified domains and amino acid residues critical for RTK activity in vivo. Moreover, genetic and bio-

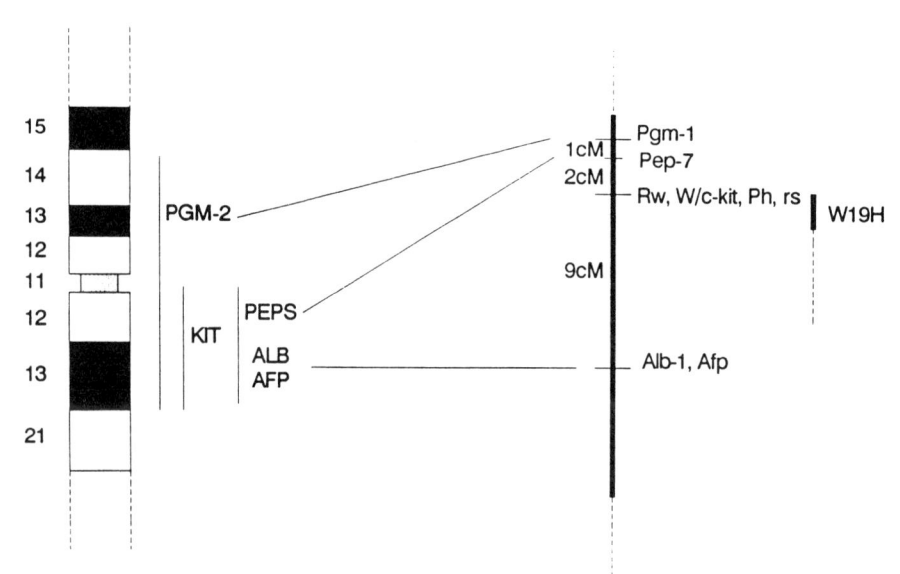

Figure 2 Chromosomal localization of *c-kit* in mouse and humans. The *c-kit* locus maps within a syntenic group of loci on human chromosome 4 and mouse chromosome 5 (Yarden et al. 1987). This syntenic group includes phosphoglucomutase 2 (PGM-2 in human, *Pgm-1* in mouse), peptidase S (PEPS in human, *Pep-7* in mouse), α-fetoprotein (AFP in human, *Afp* in mouse), and albumin (ALB in human, *Alb-1* in mouse) loci. In the mouse, this region contains a number of tightly linked developmental mutant loci including *Ramp-white* (*Rw*) (Batchelor et al. 1966; Searle and Truslove 1970), *Dominant white-spotting* (*W*) (Little 1915), *Patch* (*Ph*) (Gruneberg and Truslove 1960), and *recessive spotting* (*rs*) (Dickie 1966). In addition, a recessive lethal (Lyon and Glenister 1982) and a preimplantation lethal (Lyon et al 1984) have been mapped to this region (not shown). The mutant mouse strain W^{19H} carries a 2- to 7-cM deletion that removes *Ph* and *W*, but not *Rw* loci (Lyon et al. 1984). The *c-kit* and *pdgfra* loci are deleted in this mutant (Chabot et al. 1988; Geissler et al. 1988; Smith et al. 1991; Stephenson et al. 1991).

Table 1 Dominant pleiotropic defects induced by mutations at the W/c-kit locus

	Heterozygotes ($W^*/+$)			Homozygotes (W^*/W^*)		
	melanogenesis	hematopoiesis	fertility	melanogenesis	hematopoiesis	fertility
W^{57}	white spot	normal	normal	white patches	mild anemia	normal
W^{41}	white spot	mild anemia	normal	mostly white	severe anemia	normal
W^{44}	white spot	normal	normal	mostly white	normal	reduced
W^v	white spot	mild anemia	normal	all white	severe anemia	sterile
W^{55}	white spot	mild anemia	normal	all white	severe anemia	sterile
W	white spot	normal	normal		postimplantation lethal	
W^{37}	mostly white	normal	normal		postimplantation lethal	
W^{42}	mostly white	mild anemia	reduced		postimplantation lethal	

The phenotypic effects of various W alleles are shown in increasing order of severity. By definition, all induce dominant defects in melanogenesis, although the severity can vary. In both heterozygotes and homozygotes, an independence of pleiotropic effects is apparent for some alleles. The homozygous lethality of the W, W^{37}, and W^{42} alleles is attributable to a severe macrocytic anemia detectable at midgestation and lethal at, or near, parturition.

chemical analyses of other mouse mutants with phenotypes similar to those induced by mutation at the W/c-kit locus have revealed the identity of the ligand for the Kit receptor and possibly one other component of RTK-mediated signaling pathways.

THE W AND STEEL PHENOTYPES

Both W and Sl mutant mice are black-eyed white, with severe macrocytic anemia, mast cell deficiency, and sterility. The severity of the phenotype varies greatly among the different alleles and is more pronounced among homozygotes than heterozygotes (Table 1). Certain W and Sl alleles are homozygous lethal, with animals dying late in gestation or shortly after birth, due to severe anemia.

The similarity in pleiotropic phenotype of W and Sl mutants is striking, particularly because there is no common developmental origin of cells that give rise to melanocytes, blood cells, and germ cells. Furthermore, the W and Sl loci map on mouse chromosomes 5 and 10, respectively, so that they clearly are distinct genes. Initial insights into their mechanism of action came from in vivo cell-mixing experiments, involving either bone marrow transplantation between adult animals of different genotypes or generation of chimeric animals with aggregation chimeras. Such studies established that the defects in melanogenesis, hematopoiesis, and gametogenesis in W mutant mice are due to a cell-autonomous, intrinsic defect in the cells that gives rise to these three lineages, whereas the cellular defect in Sl mice is in the microenvironment in which these cells develop in the embryo and function in the adult animal. For example, the hematopoietic defect in W mice can be corrected by transplanting bone marrow cells from either wild-type or Sl mice (McCulloch et al. 1964; Kitamura et al. 1978; Russell 1979). In contrast, normal hematopoietic cells cannot cure the anemia or mast cell deficiency in Sl mice (McCulloch et al. 1965; Bernstein et al. 1968; Kitamura and Go 1979). Similarly, melanoblasts from wild-type neural crest survive and differentiate on normal or W/W^v skin (Mayer and Green 1968; Mayer 1970). Aggregation chimera studies between wild-type and W or Sl embryos also demonstrated that the defect in gametogenesis in both male and female W mice is not in the microenvironment, whereas the defect in Sl mutant embryos is in the microenvironment, likely in Sertoli and follicle cells of males and females, respectively (Nakayama et al. 1988; Kuroda et al. 1989).

The findings that W is allelic with c-kit receptor (Chabot et al. 1988; Geissler et al. 1988) and that Sl encodes the membrane-bound ligand for this receptor (Copeland et al. 1990; Huang et al. 1990; Zsebo et al. 1990b) have strikingly confirmed and extended these earlier biological experiments.

MOLECULAR BASIS OF *W/C-KIT* MUTANT PHENOTYPES

To date, the structure, expression, and activity of the c-*kit* gene have been analyzed in 11 independent *W* alleles. All have sustained mutations that either affect the structure of the Kit receptor or affect c-*kit* expression (Table 2; Figs. 3 and 4).

W/c-*kit* regulatory mutations

A number of *W* alleles have been characterized that result in decreased levels of expression of c-*kit* protein (Table 2; Fig. 3). Analyses of these mutants are at an early stage, but they promise to be informative with regard to defining the mechanisms that regulate c-*kit* expression.

The c-*kit* gene is deleted in the W^{19H} mutation (Chabot et al. 1988) and rearranged in W^x and W^{44} mutants (Geissler et al. 1988). Although not yet defined, the W^{44} mutation appears to be an insertion within the c-*kit* locus that affects transcription and/or stability of c-*kit* mRNA. A third mutant, *W*, results in a 234-nucleotide in-frame deletion within the c-*kit* polypeptide that includes the transmembrane domain and part of

Table 2 *W* mutants bear loss-of-function mutations in the c-*kit* RTK

	c-*kit* expression	c-*kit* kinase activity	Mutation
Wild-type mice	++++	++++	–
Mild dominance, homozygous viable			
W^{44}	++	+	genomic rearrangement
W^{57}	++	++	?
Mild dominance, homozygous lethal			
W	–	–	splice donor mutation
W^{49}	–	–	?
Moderate dominance, homozygous viable			
W^{41}	++++	++	Val-831→Met
W^{55}	++++	+	Thr-660→Met
W^v	++++	+	Thr-660→Met
W^{39}	++++	+	Met-623→Iso
Strong dominance, homozygous lethal			
W^{37}	++++	–	Glu-582→Lys
W^{42}	++++	–	Asp-790→Asn

Different classes of *W* mutant phenotypes are compared with mutations detected in c-*kit* expression and/or function. The expression and kinase activity data relate to c-*kit* proteins immunoprecipitated from in vitro cultures of mast cells homozygous for each given *W* allele.

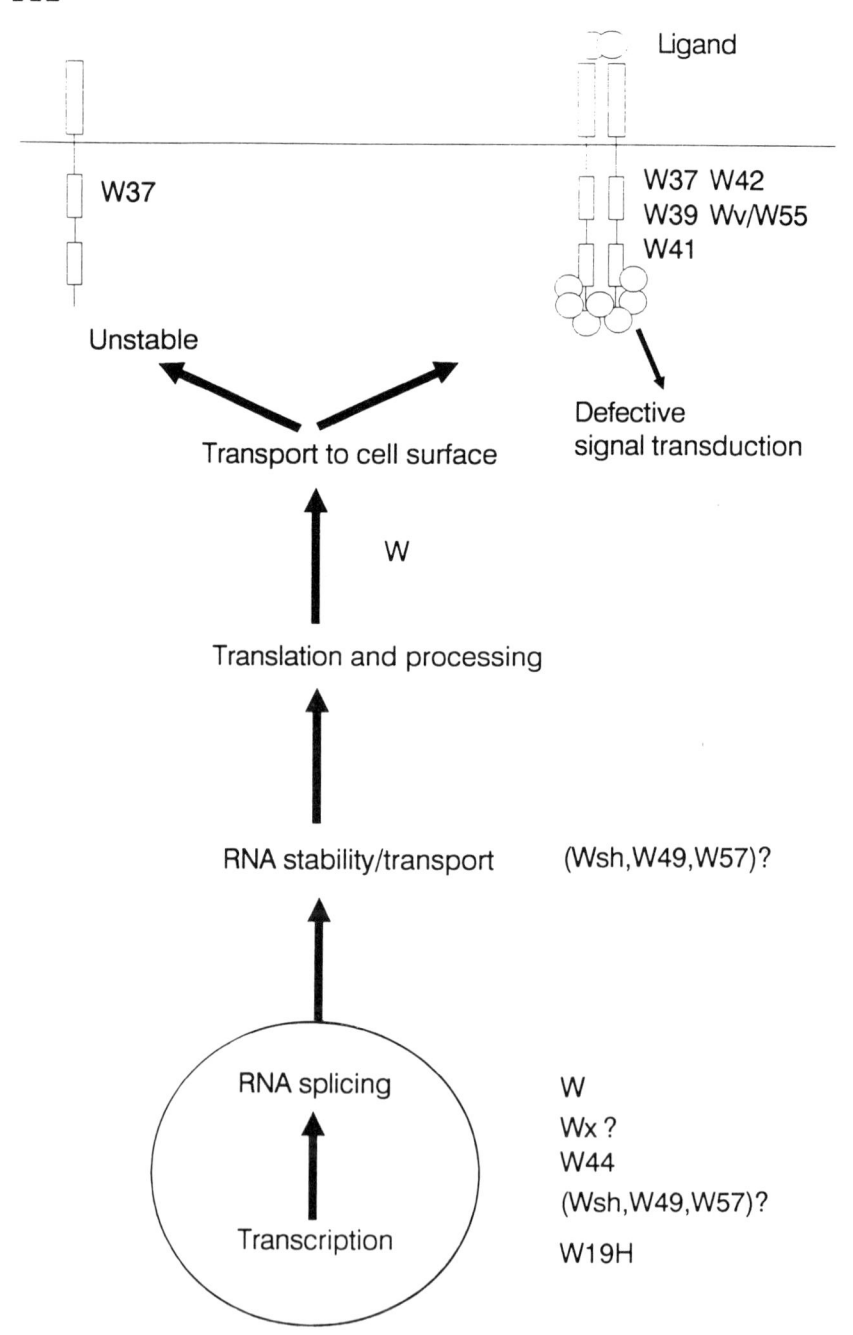

Figure 3 Biochemical defects in various *W* mutants. The known or likely biochemical defects in the various *W* mutant alleles characterized thus far are shown. The exact nature of the molecular defects in W^{sh}, W^{49}, and W^{57} have not been precisely defined, but they affect the levels of c-*kit* transcripts in mast cells, either as the result of a primary defect in transcription, RNA processing, or RNA stability. The point mutations in the W^{37}, W^{42}, W^{39}, W^{v}/W^{55}, and W^{41} alleles are shown in Fig. 4.

the kinase domain (Nocka et al. 1990b), confirming the prediction based on phenotypic analyses that W is a null allele (Lyon et al. 1984). This inframe deletion is the consequence of a single basic substitution at the 5'-splice donor site of the exon that encodes the transmembrane domain (Hayashi et al. 1991).

Quantitative reductions in the levels of immunoprecipitable c-*kit* proteins have been detected in mast cells derived from animals homozygous for W^{49} and W^{57} mutations (Table 2) (Reith et al. 1990). Although the precise nature of these mutations has yet to be defined, preliminary studies suggest that the W^{57} mutation affects c-*kit* transcription and/or mRNA stability (M. Kluppel et al., unpubl.).

At least two W alleles affect processing or stability of the Kit receptor. The W mutant form of c-*kit* does not undergo normal posttranslational processing and cannot be detected at the cell surface (Nocka et al.

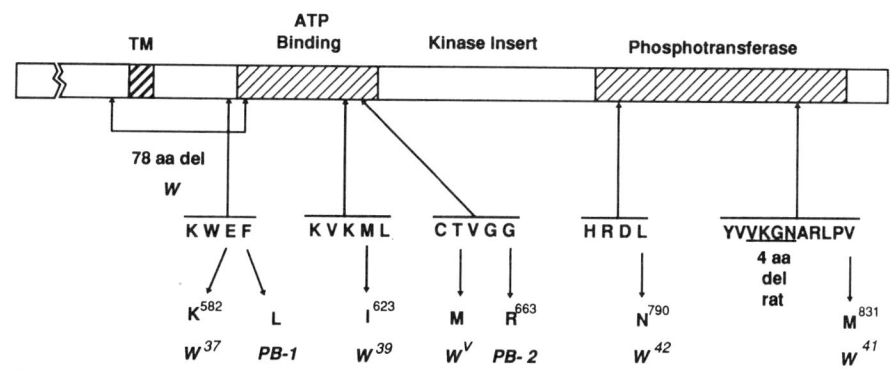

Figure 4 Mutations in the c-*kit* gene in mice, rats, and humans. Locations of point mutations and deletions detected in c-*kit* from various W mouse mutants, the Ws rat mutant, and humans with piebaldism. The point substitutions defined in the W^{42} and W^{41} alleles lie within the phosphotransferase domain, and point substitutions in W^{55} and W^v mutants are within highly conserved residues of the nucleotide binding region (Nocka et al. 1990b; Reith et al. 1990; Tan et al. 1990). The definition of the same Thr-660→Val substitution in two independent alleles, W^v and W^{55}, is consistent with the identical mutant phenotypes of these animals (see Table 1). The W^{37} substitution (Nocka et al. 1990b; Reith et al. 1990) lies outside the essential ATP-binding domain but within a region that is necessary for optimal kinase activity. The W^{39} mutation M→I at position 623 (M. Sherwood and A, Bernstein, unpubl.) is immediately downstream from the lysine at position 622 that is thought to be part of the ATP-binding site in the kinase domain. The original W mutation results in an internal deletion of 78 amino acids that includes the transmembrane domain (TM) and amino-terminal regions of the ATP-binding domain (Hayashi et al. 1991). The point mutations in two individuals with piebaldism, PB-1 and PB-2, map close to, but are distinct from, the W^{37} and W^v mutations (Giebel and Spritz 1991). The Ws rat has a 12-bp inframe deletion that removes four highly conserved amino acid residues (VKGN) in the carboxy-terminal end of the phosphotransferase domain (Tsujimura et al. 1991). Naturally occurring mutations within the extracellular or carboxy-terminal domains, or the kinase insert, of c-*kit* have yet to be identified.

1990b), consistent with the absence of the transmembrane domain in this mutant. The W^{37} allele bears a Glu→Lys point substitution within the intracellular domain that confers an instability to the 160-kD form of the Kit receptor expressed on the cell surface (Nocka et al. 1990; Reith et al. 1990b).

W/c-kit structural mutations

The second class of W/c-kit mutants is characterized by the presence of normal levels of immunoprecipitable Kit protein with qualitative defects in Kit autophosphorylation as measured by in vitro assays (Table 2; Figs. 3 and 4). To date, five W alleles in this class have been found to bear point substitutions of amino acids within the kinase domain of c-kit that are highly conserved within the family of protein tyrosine kinases (Table 2; Fig. 3) (Nocka et al. 1990b; Reith et al. 1990; Tan et al. 1990).

In some cases, these observations have confirmed an in vivo role for residues previously identified by in vitro studies as essential for kinase activity. The Asp-790→Asn mutation in the W^{42} mutant form of c-kit abolishes in vitro kinase activity (Tan et al. 1990) and was previously identified as a residue critical for kinase activity by in vitro mutagenic studies of the cytoplasmic tyrosine kinase *fps* (Moran et al. 1988). In contrast, the Glu-582→Lys W^{37} point substitution defines a previously unidentified highly conserved amino acid residue as essential for c-kit kinase activity (Nocka et al. 1990b; Reith et al. 1990), a result confirmed by in vitro mutagenic studies of recombinant c-kit molecules (Reith et al. 1991). Similarly, the point mutations in the W^{41} and W^{55}/W^v alleles (Nocka et al. 1990b; Reith et al. 1990) have identified residues that are necessary for normal levels of c-kit kinase activity.

The molecular definition of other alleles in the large collection of W mutants will undoubtedly reveal additional domains and amino acids essential for normal functioning of the Kit receptor. However, it should be noted that W mutations have all been identified on the basis of their ability to confer dominant white-spotting phenotypes and thus may represent only a subset of possible functional mutations in c-kit.

Dominant pleiotropic phenotypes induced by mutations at the W/c-kit locus

The molecular definition of W/c-kit mutations also provides a basis for understanding the mechanisms underlying the variable phenotypic defects induced by different W alleles (Table 2). First, alleles that completely abolish kit kinase as the result either of deletion or of point mutations are homozygous lethal (Table 2), whereas alleles that retain residual levels of kit kinase are homozygous viable (Table 2). Second, by comparing the phenotype of different W mutations with their molecular defects, it is apparent that mutations resulting in quantitative decreases in the level of c-kit protein confer mildly dominant defects, often only affecting coat color (Table 2). In contrast, point substitutions that affect kinase activity, but not c-*kit* protein levels, confer more strongly dominant defects, either affecting one lineage severely or affecting two or more lineages (Table 2). Phenotypes in this latter class are consistent with models of dominant-negative mutations (Kacser and Burns 1981; Herskowitz 1987) and can be understood in terms of the known mechanism of action of RTKs.

Receptor dimerization in response to ligand binding is believed to be a key step in the signal transduction process, resulting in receptor autophosphorylation and the phosphorylation of specific intracellular protein targets (Ullrich and Schlessinger 1990). Kinase-defective Kit molecules would sequester normal receptor into nonfunctional heterodimers, markedly reducing the efficiency of signal transduction and resulting in a more strongly dominant loss-of-function phenotype than a regulatory mutation, which solely affects the levels of Kit receptor expression. The identification of point substitutions that induce such dominant-negative phenotypes has important implications in the design of strategies to study the in vivo roles of other RTKs for which germ-line mouse mutants are not yet available (see below).

Homozygous lethal *W* alleles bear either regulatory or structural mutations that abolish c-*kit* kinase activity, indicating that the c-*kit* signaling pathway is essential for viability, probably a consequence of the virtual lack of erythropoiesis in such homozygous embryos. The homozygous viable *W* mutants are characterized either by quantitative reductions in levels of a normal c-*kit* protein, or structural mutations that reduce but do not abolish c-*kit* in vitro kinase activity. For some alleles (W^v/W^{55}, W^{39}), all three lineages are affected to similar extents, whereas others (W^{44}, W^{41}, and W^{57}) exhibit selective pleiotropic effects in the homozygous state, as do W^{37} heterozygous animals (Table 1). The molecular basis of such differences in severity is not evident at the present time. Residual levels of c-*kit* kinase activity may be sufficient for normal development of some lineages, but not others. More interestingly, the mutations may act in a cell-specific manner. The definition of these mutants, and the availability of in vitro culture systems for cell types that are affected by *W* mutations, will facilitate a greater understanding of the molecular mechanisms underlying these *W* phenotypes.

THE KIT SIGNAL TRANSDUCTION PATHWAY

RTKs represent only one component of a signaling pathway, interacting with both ligands and intracellular substrates to evoke a proliferative response. Biochemical characterization of proteins acting in the same pathway as RTKs has been carried out by identification of receptor-associated proteins and substrates that form a physical complex with ligand-activated receptors (for review, see Ullrich and Schlessinger 1990; Koch et al. 1991). The availability of mutant mouse strains with phenotypes identical with or overlapping those of *W* mice (Lyon and Searle 1989) suggests that such mutants may affect c-*kit* biosynthesis or are components of the Kit signaling pathways. Thus, it may be possible to undertake a genetic approach to identify proteins that lie either upstream or downstream from the Kit receptor. This approach has resulted in the identification of *steel* as the gene that encodes the Kit ligand and *microphthalmia* (*mi*), the product of which appears to act downstream from c-*kit*.

Sl encodes the membrane-bound ligand for the Kit receptor

As discussed above, mutations at the *Steel* (*Sl*) locus on chromosome 10 confer mutant phenotypes identical to those induced by *W* mutations, affecting the hematopoietic, melanogenic, and germ-cell lineages (Sarvella and Russell 1956; Russell 1979; Silvers 1979). However, the cellular defect in these mice is in the microenvironment in which these three lineages develop in the embryo and function in the adult. The finding that *W* is allelic with c-*kit* suggested that *Sl* might encode the ligand or growth factor that recognizes and activates the Kit receptor (Chabot et al. 1988). This possibility was further supported by two findings. First, normal mast cells, but not mast cells derived from W/W^v mice, can grow on a monolayer of mouse embryo fibroblasts (Fujita et al. 1988a,b; Jarboe and Huff 1989). Second, normal mouse embryo fibroblasts, but not fibroblasts derived from Sl/Sl^d mice, can support the growth of normal mast cells (Fujita et al. 1989; Jarboe and Huff 1989). Taken together, these results are entirely consistent with the notion that the product of the *Sl* locus, made by fibroblasts, is a growth factor that acts via the Kit receptor expressed on mast cells.

Subsequently, a number of groups identified mast cell growth factors (MGF) produced by normal or W/W^v fibroblasts, but not Sl/Sl^d fibroblasts, that bind to the Kit receptor (Flanagan and Leder 1990; Huang et al. 1990; Nocka et al. 1990a; Williams et al. 1990). Cloning of these activities led to the isolation of cDNA clones that encode both soluble and membrane-bound forms of MGF (Anderson et al. 1990; Huang et al. 1990). Independently, a stem-cell growth factor, active on primitive hematopoietic precursors and therefore designated stem cell factor (SCF), was purified from Buffalo rat liver conditioned medium and found to be encoded by the same gene (Martin et al. 1990; Zsebo et al. 1990a,b). The gene for these factors maps to the *Sl* locus and is deleted or rearranged in a number of independent *Sl* alleles (Copeland et al. 1990; Huang et al. 1990; Zsebo et al. 1990b; Brannan et al. 1991; Flanagan et al. 1991), indicating that the *Sl* locus does indeed encode a ligand for the *W*/c-*kit* receptor. The factor encoded by the *Sl* locus has been termed MGF, SCF, and KL (Kit ligand) and is referred to here as

Steel factor. Alternative splicing of the *Sl* transcript results in both secreted and membrane-bound forms of Steel factor (Anderson et al. 1990; Flanagan and Leder 1990;).

Like mildly dominant *W* mutants, a reduction in *Steel* protein levels appears sufficient to induce mildly dominant phenotypes, presumably as a result of decreased signaling through the Kit receptor. Deletion of *Steel* results in homozygous lethal phenotypes, similar to those *W* mutations that abolish c-*kit* kinase activity. As with the *W*/c-*kit* locus, the molecular definition of more subtle *Sl* mutations is likely to be most informative with regard to understanding the mechanisms of Steel factor biosynthesis and bioactivity. For example, the mild Sl^d allele is a deletion of the transmembrane domain resulting in the expression of only the soluble form of Steel factor (Brannan et al. 1991; Flanagan et al. 1991). Thus, it appears that Steel factor is normally presented to the Kit receptor as a membrane-bound molecule, a conclusion consistent with in situ analyses of c-*kit* and *Sl* expression described below and earlier biological experiments that suggested that *Sl* controlled the local microenvironment of cells affected by the *W* locus. Further analysis of Sl^{panB} and Sl^{17H} alleles, for which genomic rearrangements within the *Sl* locus have not been detected by Southern blot analyses, will be particularly useful in further defining the mechanism of action of Steel factor and the mechanisms that regulate expression of the *Sl* gene.

Cytoplasmic substrates of the ligand-activated Kit receptor

The binding of ligand to RTKs, such as the Kit receptor, initiates an array of protein-protein interactions between receptors and downstream signaling proteins that culminates in a proliferative or developmental response. Ligand binding induces receptor dimerization and autophosphorylation at specific tyrosine residues within the cytoplasmic domain (Ullrich and Schlessinger 1990). The phosphorylated forms of these receptors can then bind and phosphorylate a unique array of cytoplasmic signaling proteins to evoke a cellular response. These signaling proteins include phosphatidylinositol (PI) 3'-kinase, phospholipase (PLC)-γ1, Ras GTPase activating protein (GAP), and members of the *src* family of tyrosine kinases (Ullrich and Schlessinger 1990; Koch et al. 1991).

Activation of the Kit receptor by the binding of Steel factor rapidly induces Kit autophosphorylation and its binding to PI3'-kinase and PLCγ1. These reactions are either completely or partially blocked with Kit receptors carrying severe (e.g., W^{37}, W^{42}) or mild (e.g., W^v/W^{55}) mutations, respectively (Reith et al. 1991; Rottapel et al. 1991). Different RTKs, even within the structurally related Kit, CSF-1R, PDGFR subfamily of receptors, interact with an overlapping, but unique, array of downstream signaling proteins. Thus, the β-PDGFR binds to PI3'-kinase, GAP, and PLCγ1; the CSF-1R only binds to PI3'-kinase, whereas, as discussed above, Kit binds to PI3'-kinase and PLCγ1. This specificity of interactions, together with the differences in transcriptional regulation of these receptors and the differences in their cognate ligands, may together account for the unique biological roles of each member of this subfamily of receptors.

How do activated receptors interact with proteins such as PI3'-kinase and GAP? Experiments with receptor deletion mutants and receptor fragments expressed in bacteria suggest that tyrosine phosphorylation of the noncatalytic kinase insert (KI) regions of the PDGFR (Kazlauskaus and Cooper 1989; Yu et al. 1991) and the CSF-1R and the Kit receptors (Rottapel et al. 1991) is necessary for binding to molecules such as PI3'-kinase. Thus, tyrosine autophosphorylation of the Kit receptor acts as a switch that creates active binding sites for downstream signaling proteins.

The interactions of these proteins with receptors such as the Kit receptor are mediated by two copies of a noncatalytic domain originally defined by a consensus sequence present in the Src tyrosine kinase and hence called SH2 for Src homology 2 (for review, see Koch et al. 1991). Bacterially expressed SH2 domains derived from PLCγ1 and the p85α regulatory subunit of PI3'-kinase associate with the phosphorylated Kit receptor (C.J. McGlade et al., in prep.). The phenotypes associated with structural mutations at the W locus therefore result from the partial or complete block in the formation of these active binding sites for downstream signaling proteins as the result of alterations in the intrinsic kinase activity of the Kit receptor. Interestingly, none of the W mutants analyzed to date results from alterations within the KI region.

microphthalmia: A locus required for Kit and FMS function

The phenotype of animals bearing mutations at the *mi* locus (chromosome 6) (Hertwig 1942) shares some similarities with those of W and Sl mutants. Homozygosity for the *mi* mutation results in defects in pigmentation, eye development, bone resorption, and mast cells (Silvers 1979; Lyon and Searle 1989). Like mutations at the W/c-kit locus, the *mi* hematopoietic defect is intrinsic to stem cells of this lineage (Walker 1975; Marks 1984). Moreover, although *mi/mi* mast cells exhibit normal growth in response to IL-3, they are defective in fibroblast-dependent growth (Dubreuil et al. 1990; Ebi et al. 1990), suggesting that they might be unable to respond to Steel factor made by fibroblasts. These observations, together with the fact that the phenotypic abnormalities in *mi* mutant mice resemble those of W and Sl mice (pigmentation, mast cell deficiencies) and the *osteopetrotic (op)* mouse that is mutated in the gene for CSF-1 (Yoshida et al. 1990), suggest that *mi* controls a common downstream step in the Kit and FMS signaling pathways.

In support of this hypothesis, the defects conferred by both W and

Steel mutations on mast cell viability in the fibroblast co-culture assay can be overcome by ectopic expression of c-*fms*, a RTK closely related to c-*kit* (Dubreuil et al. 1990). In contrast, the *mi* defect cannot be overcome by expression of c-*fms*, an observation consistent with the idea that the *mi* gene product(s) is a component common to both MGF/c-*kit* and CSF-1/c-*fms* signal transduction pathways in mast cells (Dubreuil et al. 1990).

THE IN VIVO FUNCTIONS OF THE KIT SIGNAL TRANSDUCTION PATHWAY

As discussed below, the Kit-Steel interaction clearly acts as a proliferative signal to cells within the hematopoietic stem-cell hierarchy. Both the ongoing deficiency in red blood cells and mast cells in adult *W* and *Sl* mutant mice and the defects in oocyte maturation in mild *W* and *Sl* mutants suggest that the Kit pathway is tightly coupled to the mitogenic response in vivo. The Kit receptor also may play a role in cell migration during development. The cellular defects in *W* and *Sl* mice first become apparent in embryogenesis when germ cells are migrating from the yolk sac along the genital ridges to the gonads, when melanoblasts migrate from the neural tube to the dermis, and when hematopoiesis shifts from the yolk sac blood islands to fetal liver. Three additional observations also support a role for the Kit signaling pathway in cell migration. First, as noted above, Steel factor is normally presented to the Kit receptor as a membrane-bound ligand so that Steel expression on stromal cells may provide localized cues for cell migration and cell chemotaxis. Second, in situ RNA hybridization analysis has shown that Steel is expressed in cells that line the migratory routes of germ cells, melanoblasts, and hematopoietic stem cells (Matsui et al. 1990). Third, cells expressing c-*kit* adhere to fibroblasts that are synthesizing the membrane-bound form of Steel factor (Kaneko et al. 1991). In addition to a possible role of the Kit-Steel pathway in migration, it is clear, as discussed below, that Steel activation of the Kit receptor induces a strong proliferative response to cells.

Hematopoiesis

During embryonic development, hematopoiesis first takes place in the yolk sac blood islands between days 7 and 11 of gestation. The fetal liver then becomes the major site of hematopoiesis between days 11.5 and 18, probably as the result of migration of stem cells from the yolk sac to the fetal liver. Hematopoiesis then shifts to the bone marrow from day 16 into adult life. Thus, the hematopoietic system undergoes two major migratory waves during embryogenesis, suggesting, as discussed above, the possibility that the Kit pathway is important to this process. *W* and *Sl* mutant mice exhibit ongoing hematopoietic defects in adult mice, indicating that the Kit pathway has an ongoing function in the adult. What cell types within the hematopoietic system utilize the Kit receptor, and what are the consequences of activating the Kit receptor by the addition of its ligand, Steel factor? The hematopoietic system is composed of a cellular hierarchy, ranging from pluripotent stem cells with extensive developmental and proliferative capacity, to committed progenitor cells that are more limited in their developmental options and proliferative capacity, to fully differentiated mature blood cells. As *W* and *Sl* mice are both anemic and mast-cell-deficient, it is clear that the Kit pathway must be utilized by progenitor cells that give rise to red blood cells and mast cells.

Two approaches have been taken to identify these cells. First, cells expressing the Kit receptor have been identified and partially characterized in hematopoietic cell populations. Second, cells responding to Steel factor in unfractionated and fractionated cell populations have been identified. By sorting bone marrow cells with monoclonal antibodies directed to a variety of surface antigens (CD33, CD34) or to the extracellular domain of the Kit receptor, it appears that Kit is expressed on a small fraction (4%) of adult human (Ashman et al. 1991) or mouse (Ogawa et al. 1991; Papayannopoulou et al. 1991) marrow cells. The Kit$^+$ cells are distributed among cells that either express or fail to express certain antigens associated with myeloid or lymphoid lineages (Ashman et al. 1991; Bernstein et al. 1991) and are highly enriched for hematopoietic progenitor cells that are able to form colonies in vitro in response to various growth factors (Papayannopoulou et al. 1991). The addition of Steel factor to either unfractionated or fractionated hematopoietic cell populations stimulates the growth of small, undifferentiated "blast-like" colonies. Most strikingly, in combination with other cytokines, including EPO, IL-3, G-CSF, and GM-CSF, Steel factor acts synergistically as a potent comitogen to stimulate directly the growth of a variety of committed progenitor cells (Bernstein et al. 1991; Broxmeyer et al. 1991; McNiece et al. 1991a,b; Metcalf and Nicola 1991; Migliaccio et al. 1991; Olivieri et al. 1991). These studies clearly demonstrate functional expression of the Kit receptor on a variety of different cell types within the hematopoietic stem cell hierarchy. Cells expressing the Kit receptor in murine bone marrow are also capable of reconstituting the hematopoietic system of lethally irradiated mice for at least 5 weeks, suggesting that the earliest cells in the stem cell hierarchy may utilize this signaling pathway (Ogawa et al. 1991).

Melanogenesis

The pigmentation defect of *W* mice results from intrinsic deficiencies in neural-crest-derived melanoblasts. Normally, melanoblasts migrate to populate the dermis, and subsequently the epidermis, around 13–14 days gestation. Grafting studies demonstrated that W melanoblasts either never enter the skin or die soon after, and that the Sl mutation acts within the dermis and epidermis to prevent melanoblast differentiation and/or melanoblast or melanocyte survival. The detection of c-kit transcripts in melanocyte precursors in early mouse embryos (Orr-Urtreger et al. 1990; Manova and Bachvarova 1991) is consistent with a function for c-kit at this stage in melanocyte development, as is the presence of steel transcripts, in the regions of presumed migration of melanoblasts in the embryo (Matsui et al. 1990). Recently, monoclonal antibodies against the c-kit extracellular domain have been used as antagonists to address directly c-kit functions during melanogenesis (Nishikawa et al. 1990). These experiments indicate that c-kit is essential during the migration of melanoblasts to colonize hair follicles. However, whether c-kit function is required for proliferation of melanocytes or colonization of hair follicles at this stage is unknown. This approach has also shown that c-kit function is additionally required for melanocyte activation postnatally, a result consistent with the expression of c-kit in normal mouse melanocytes (Nocka et al. 1989).

Gametogenesis

Like melanogenesis, germ cell development involves extensive proliferation coupled with cellular migration during embryogenesis. In normal mice, germ cells are first detected around 8 days gestation in the yolk sac splanchopleure, allantoic mesoderm, and caudal primitive streak. Between 9 and 12 days gestation, they proliferate rapidly as they migrate toward the genital ridge (Mintz and Russell 1957). In W or W^v homozygotes, normal numbers of germ cells are present at 8 days gestation, but they subsequently fail to proliferate and are retarded in migration (Mintz and Russell 1957). c-kit transcripts have yet to be detected in primordial germ cells during their migration to the genital ridge, but expression has been found in primordial germ cells within the genital ridge at 12.5 days gestation (Orr-Urtreger et al. 1990). Sl expression has been detected both in mesodermal cells along the migratory path of germ cells from 9 days gestation onward, and within the genital ridge (Matsui et al. 1990), consistent with a function of the c-kit signaling pathway in germ cell development. Moreover, c-kit expression in both germ cells and supporting tissues of testis and ovary in juvenile and adult mice (Manova et al. 1990; Orr-Urtreger et al. 1990; Manova and Bachvarova 1991; Motro et al. 1991) suggests that the Kit signaling pathway may also function at later stages of germ cell development, a conclusion further supported by expression of steel in Sertoli cells in adult testis and follicle cells in the adult ovary (Keshet et al. 1991; Motro et al. 1991).

In support of the hypothesis that the Kit signaling pathway is important to gametogenesis, two groups have recently shown that the survival of primordial germ cells in culture is enhanced by Steel factor (Dolci et al. 1991; Godin et al. 1991). Reflecting the sterility associated with the Sl^d mutation, which does not make the membrane-bound form of Steel factor, the soluble form of Steel factor is not effective in promoting primordial germ cell survival (Dolci et al. 1991). Furthermore, by itself, Steel factor does not act as a mitogenic or chemotrophic factor for primordial germ cells, a conclusion consistent with much earlier findings that in Sl/Sl^d mice that express only the soluble form of Steel factor, primordial germ cells appear and migrate to the gonads but do not survive (McCoshen and McCallion 1975).

CONTIGUOUS PATTERNS OF C-*KIT* AND *STEEL* EXPRESSION DURING EMBRYOGENESIS AND IN THE ADULT

The broad diversity of cells expressing c-*kit* and *steel* transcends both common developmental origin and function. For example, c-*kit* is expressed in cells derived from ectoderm, mesoderm, and endoderm. In the neuroectoderm, c-*kit* is expressed in both neurons and glial cells. With respect to function, c-*kit* is expressed in both limbic (e.g., hypothalamus and hippocampus) and sensory motor regions (e.g., neocortex), in deep cerebellar neurons that receive synaptic input from inhibitory neurons (Purkinje cells), and in CA3 pyramidal neurons that receive synaptic input from excitatory neurons in the dentate gyrus (Motro et al. 1991).

In situ RNA hybridization of consecutive tissue sections has revealed the strikingly complementary patterns of expression of the c-*kit* and *steel* genes and the close physical proximity between cells expressing c-*kit* (frequently seen as individual positive cells) and the band of cells expressing *steel*. The RNA in situ hybridization analyses support earlier conclusions that *steel* acts by affecting the microenvironment in which other cells develop, migrate, and proliferate. This conclusion was initially based on the phenotype of chimeras, established at the embryonic stage (Nakayama et al. 1988), or by the transplantation of splenic fragments between wild-type and *Sl* mutants (Harrison and Russell 1972). A similar conclusion has been reached from the observations that the protein product of the *steel* locus is membrane-bound (Anderson et al. 1990; Flanagan and Leder 1990) and that the mild *Steel* allele Sl^d does not make the membrane-bound form of Steel factor (Brannan et al. 1991; Flanagan et al. 1991). Taken together, these observations, derived from three entirely different experimental approaches, all suggest that activation of the Kit receptor by binding of the Steel factor normally occurs via cell-cell interactions in which the membrane-bound form of Steel factor on one cell binds to the extracellular domain of the Kit receptor on a neighboring cell. This binding then initiates the series of intracellular signaling events described above.

The complementary patterns of c-*kit* and *steel* expression are observed in both embryos and adult animals (Keshet et al. 1991; Motro et al. 1991). These observations suggest that the Kit pathway has an ongoing function in the adult, a conclusion consistent with the phenotypic analyses of viable *W* and *Sl* mutants. For example, W/W^v adult mice are severely anemic and have reduced numbers of erythroid progenitor cells and defective mast cell proliferation. Similarly, oocytes are present in Sl/Sl^t females, but they fail to mature (Kuroda et al. 1988). The observation that follicle cells express high levels of *steel* in the adult female and that oocytes express c-*kit* is consistent with a role for the Kit pathway in oocyte maturation. Similarly, the ongoing expression of *steel* in Sertoli cells in the adult testes, together with the impaired differentiation of germ cells in *Sl/+* adult males (Nishimune et al. 1984) and the ability of injected anti-c-*kit* monoclonal antibodies to block spermatogonial maturation (Yoshinaga et al. 1991), suggests that activation of the Kit receptor is required for spermatogenesis.

The highly contiguous patterns of expression of c-*kit* and *steel* in a broad array of distinct anatomical sites make it likely that the c-*kit* signaling pathway has important biological functions in these diverse tissues and cell types. Nevertheless, mutations at the *W* or *Sl* loci do not have any obvious phenotypic consequences outside of their profound effect on melanogenesis, hematopoiesis, and gametogenesis. For example, no detectable change in the number or organization of cells in the CNS that express c-*kit* is observed in *W* or *Sl* mutants (Motro et al. 1991), suggesting that the Kit signaling pathway is not essential for cellular proliferation and organogenesis in those sites such as the brain that are not phenotypically affected by mutation at the *W* or *Steel* loci.

There are several possible explanations for these results. First, phenotypic abnormalities may not be evident because they are subtle. Second, the c-*kit* signaling pathway may be redundant with other signal transduction pathways, particularly in the brain, where a large number of both receptor and nonreceptor tyrosine kinases are highly expressed (see, e.g., Lai and Lemke 1991). Functional redundancy with another RTK pathway could also explain the absence of placental defects in *Sl/Sl* embryos. Third, the absence of a functional c-*kit* pathway may affect postdevelopmental and postmitotic cell signaling events, causing phenotypic abnormalities not readily detectable by morphology alone. For example, in the brain, the Kit pathways may participate in the processes of axonal pathfinding and synaptogenesis.

The *Ph-W-Rw-rs* gene cluster

In addition to the *W* locus, mutations in three other loci on mouse chromosome 5, *Patch* (*Ph*), *Rump-white* (*Rw*), and *recessive spotting* (*rs*), lead to white-spotting as the result of effects on developing melanoblasts (see Fig. 2). More than 50% of *Ph/Ph* embryos homozygous for the *Patch* (*Ph*) mutation display gross anatomical abnormalities and die midway through gestation. Embryos homozygous for the *Rump-white* (*Rw/Rw*) also die in utero for unknown reasons. In addition to their common effect on coat color, these loci are extremely closely linked genetically. Although complementation analysis suggested that *Ph* is distinct from *W*, both *W* and *Ph* (but not *Rw*) are deleted in the deletion defined by the W^{19H} allele (Smith et al. 1991; Stephenson et al. 1991)), suggesting close physical linkage of the *W* and *Ph* loci.

The close linkage between these loci, as well as the common effects on melanocyte development that result from mutations in either membrane of this gene cluster, suggests some common evolutionary origin for these genes. Recently, two groups independently demonstrated that the murine gene for the α-PDGF receptor (*Pdgfra*), a member of the Kit/PDGFR/FMS subfamily of RTKs, is closely linked to *W/c-kit* on mouse chromosome 5 and can be colocalized on a common 630-kbp DNA segment that includes c-*kit* (Smith et al. 1991; Stephenson et al. 1991). Furthermore, like c-*kit*, *Pdgfra* is deleted in the W^{19H} allele. In addition, although the c-*kit* gene appears to be intact in DNA from *Ph* mice, the *Pdgfra* gene is deleted, suggesting that *Pdgfra* and *Ph* may be allelic.

These observations suggest that a primordial member of this subfamily of RTKs underwent a gene duplication event resulting in the close physical configuration of c-*kit* and *Pdgfra*. It will be interesting to determine whether *Rw* is allelic with yet another member of this subfamily of receptors.

Finally, it is of interest that two other members of this receptor subfamily, *pdgfrb* and c-*fms*, are physically linked in a head-to-tail tandem array within 500 bp of each other on human chromosome 5 (Roberts et al. 1988; Sherr 1990). During evolution, there must therefore have been either a dispersion event of a single primordial gene followed by two independent duplication (or triplication) events in situ or a gene amplification event followed by dispersion to other chromosomes.

W/c-kit mutations in other species

The high mutation rate at the murine *W* locus (Schlager and Dickie 1971), together with the easily discernible and dominant coat color defects associated with mutations at this loci, suggests that mutations within c-*kit* should exist in other mammals. The *Ws* mutation appeared spontaneously in the BN strain of rats. Heterozygote *Ws/+* rats have diluted coat color and ventral white spot with mild anemia that resembles $W^v/+$ mice, whereas presumptive *Ws/Ws* animals are born white with a severe macrocytic anemia (Niwa et al. 1991). The c-*kit* gene in these mutant rats has sustained a 12-bp in-frame deletion that results in a deletion of four highly conserved amino acids at positions 826–829 immediately downstream from a presumptive tyrosine phosphorylation site in the Kit receptor (Fig. 4) (Tsujimura et al. 1991).

Piebaldism is an autosomal dominant disorder in humans characterized by white patches of skin and hair, usually on the forehead, abdomen, and ventral chest. This phenotype, together with the cytogenetic data implicating the long arm of chromosome 4 where the human c-*kit* gene maps, suggested that human piebaldism might result from mutations of the c-*kit* gene. DNA sequencing and Southern blot analysis of the c-*kit* gene from four unrelated families with piebaldism has demonstrated deletion, point mutations, and a frameshift mutation within the cytoplasmic kinase domain of the Kit receptor, suggesting that this disorder is the human equivalent of the dominant whitespotting *W* locus in the mouse (Giebel and Spritz 1991; R. Fleischman, pers. comm.; A. Bernstein et al., unpubl.). Interestingly, the point mutations described so far in piebald patients lie close to, but are distinct from, the point substitutions observed in *W* mutants (Fig. 4). These observations suggest that mutations within a large number of amino acid residues can lead to the dominant-negative *W* phenotype, an observation that might explain the high mutation rate of this locus in mice.

New members of the c-*kit* family of RTKs

It is clear that RTKs play a central role in cell-cell communication during development and differentiation and that this family of receptors has undergone extensive growth and evolution, reflecting the complexity and diversity of specialized cell types in multicellular organisms. Several new members of the c-*kit*/c-*fms*/*pdgfr* family of RTKs have recently been isolated either by PCR-based strategies or by screening libraries under conditions of reduced stringency (Shibuya et al. 1990; Matthews et al. 1991; Rosnet et al. 1991). The role of these novel members of the c-*kit* gene family in vivo is not yet known. The analysis of *W* and *Sl* mutant mice demonstrates the power of a genetic approach to address such questions. Although naturally occurring mutations in the genes that encode these novel RTKs are not available, it is possible to generate such mutants by manipulating the mouse germ line. Gain-of-function mutations can be generated through the random insertion into the germ line of DNA constructs that result in ectopic or overexpression, whereas loss-of-function mutants can be derived by targeting the insertion of exogenous DNA sequences into a specific chromosomal site by homologous recombination (Capecchi 1989; Rossant and Joyner 1989). In addition, the dominant-negative nature of *W* structural mutants suggests that it should be possible to generate dominant-negative mutations in other RTKs by introducing point substitutions equivalent to those found in the highly conserved amino acid residues that are mutated in different *W* alleles. This strategy has been shown to work in vitro in blocking CSF-1-dependent transformation of fibroblasts by c-*fms* (A Reith et al., in prep.).

RTK SIGNALING PATHWAYS AND DEVELOPMENT

The association of loss-of-function mutations in the genes for the c-*kit* receptor or its ligand with *W* and *Sl* mutant phenotypes has revealed much about the role of this signaling pathway in mammalian development and served to illustrate the role played by receptor-ligand interactions in directing appropriate patterns of cell migration, proliferation, and differentiation during embryogenesis. The membrane-bound nature of Steel factor, the ligand for the Kit receptor, has also provided perhaps the best example of short-range positional cues acting via cell-cell contact to govern the behavior of mammalian cells. In this respect, the biology and molecular biology of the *W-Sl* gene pair are strikingly analogous to the *sevenless/bride of sevenless* (*boss*) genes in *Drosophila*, which, respectively, encode a RTK (Hafen et al. 1987) and its membrane-bound ligand (Kramer et al. 1991) that together control the fate of the R7 photoreceptor during development of the compound eye.

Cell-cell interactions play a major role both in the determination of cell fate within the early embryo and in the differentiation of specific cell types during embryogenesis and adult life. In many cases, these developmental decisions involve activation of the biochemical pathways controlled by a specific RTK. The molecular genetic analysis of the murine *W* and *Sl* loci, coupled with the biological description of the cellular defects in these mutant mice, illustrates the power of combining biological, genetic, and molecular approaches to dissect the complex steps in mammalian development.

Acknowledgments

We thank Christine Ellis, Benny Motro, and Norm Lassam for many useful discussions concerning RTK functions in development and for constructive criticisms of the manuscript. We appreciate Ken Kao's assistance in the preparation of Figure 4 and Susan Varga's help with the preparation of the manuscript. A.D.R. acknowledges receipt of a NATO postdoctoral fellowship during part of this work. A.B. is a Howard Hughes International Scholar. Work from the authors' laboratory was supported by grants from the National Institutes of Health and the National Cancer Institute of Canada.

References

Anderson, D.M., S.D. Lyman, A. Baird, J.M. Wignall, J. Eisenman C. Rauch, C.J. March, H.S. Boswell, S.D. Gimpel, D. Cosman, and D.E. Williams. 1990. Molecular cloning of mast cell growth factor, a hematopoietin that is active in both membrane bound and soluble forms. *Cell* **63**: 235.

Ashman, L.K, A.C. Cambareri, L. Bik To, R.J. Levinsky, and C.A. Juttner. 1991. Expression of the YB5.B8 antigen (c-*kit* proto-oncogene product) in normal human bone marrow. *Blood* **78**: 30.

Batchelor, A.L., R.J.S. Phillips, and A.G. Searle. 1966. A comparison of the mutagenic effectiveness of chronic neutron and γ-irradiation of mouse spermatogonia. *Mutat. Res.* **3**: 218.

Bernstein, I.D., R.G. Andrews, and K.M. Zsebo. 1991. Recombination human stem cell factor enhances the formation of colonies by CD34+ and CD34+ lin− cells, and the generation of colony-forming cell progeny from CD34+ lin− cells cultured with interleukin-3, granulocyte colony-stimulating factor, or granulocyte-macrophage colony-stimulating factor. *Blood* **77**: 2316.

Bernstein, S.E., E.S. Russell, and G.H. Keighley. 1968 Two hereditary mouse anemias (*Sl/Sl^d* and *W/W^v*) deficient in response to erythropoietin. *Ann. N.Y. Acad. Sci.* **149**: 475.

Besmer, P., J.E. Murphy, P.C. George, F. Qui, P.J. Bergold, L. Lederman, H.W. Snyder, Jr., D. Brodeur, E.E. Zuckerman, and W.D. Hardy. 1986. A new acute transforming feline retrovirus and relationship of its oncogene v-*kit* with the protein kinase gene family. *Nature* **320**: 415.

Brannan, C.I., S.D. Lyman, D.E. Williams, J. Eisenman, D.M. Anderson, D. Cosman, M.A. Bedell, N.A. Jenkins, and N.G. Copeland. 1991. Steel-*Dickie* mutation encodes c-*kit* ligand lacking transmembrane and cytoplasmic domains. *Proc. Natl. Acad. Sci.* **88**: 4671.

Broxmeyer, H.E., S. Cooper, L. Lu, G. Hangoc, D. Anderson, D. Cosman, S.D. Lyman, and D.E. Williams. 1991. Effect of murine mast cell growth factor (c-*kit* proto-oncogene ligand) on colony formation by human marrow hematopoietic progenitor cells. *Blood* **77**: 2142.

Capecchi, M.R. 1989. Altering the genome by homologous recombination. *Science* **244**: 1288.

Chabot, B., D.A. Stephenson, W.M. Chapman, P. Besmer, and A. Bernstein. 1988. The proto-oncogene c-*kit* encoding a transmembrane tyrosine kinase receptor maps to the mouse *W* locus. *Nature* **335**: 88.

Copeland, N.G., D.J. Gilbert, B.C. Cho, P.J. Donovan, N.A. Jenkins, D. Cosman, D. Anderson, S.D. Lyman, and D.E. Williams. 1990. Mast cell growth factor maps near the *Steel* locus on mouse chromosome 10 and is deleted in a number of *Steel* alleles. *Cell* **63**: 175.

Dolci, S., D.E. Williams, M.K. Ernst, J.L. Resnick, C.I. Brannan, L.F. Lock, S.D. Lyman, H.S. Boswell, and P.J. Donovan. 1991. Requirement for mast cell growth factor for primordial germ cell survival in culture. *Nature* **352**: 809.

Dickie, M.M. 1966. Private communication. *Mouse News Lett.* **35**: 31.

Dubreuil, P., L. Forrester, R. Rottapel, J. Fujita, and A. Bernstein. 1990. The c-*fms* gene complements the mitogenic defect in mast cells derived from mutant *W* mice but not *mi* (*micropthalmia*) mice. *Proc. Natl. Acad. Sci.* **88**: 2341.

Ebi, Y., T. Kasugai, Y. Seino, H. Onoue, T. Kanemoto, and Y. Kitamura. 1990. Mechanism of mast cell deficiency in mutant mice of *mi/mi* genotype: An analysis by coculture of mast cells and fibroblasts. *Blood* 75: 1247.

Flanagan, J.G. and P. Leder. 1990. The *kit* ligand: A cell surface molecule altered in Steel mutant fibroblasts. *Cell* 63: 185.

Flanagan, J.G., D.C. Chan, and P. Leder. 1991. Transmembrane form of the *Kit* ligand growth factor is determined by alternative splicing and is missing in the Sl^d mutant. *Cell* 65: 1025.

Fujita, J., H. Oncoue, Y. Ebi, H. Nakayama, and Y. Kanakura. 1989. *In vitro* duplication and *in vivo* cure of mast cell deficiency of Sl/Sl^d mutant mice by cloned 3T3 fibroblasts. *Proc. Natl. Acad. Sci.* 86: 2888.

Fujita, H., H. Nakayama, H. Onoue, Y. Ebi, Y. Kanakura, A. Kurin, and Y. Kitamura. 1988a. Failure of W/W^v mouse-derived cultured mast cells to enter S phase upon contact with NIH3T3 fibroblasts. *Blood* 72: 463.

Fujita, J., H. Nakayama, H. Onoue, Y. Kanakura, T. Nakano, H. Asai, S. Takeda, T. Honjo, and Y. Kitamura. 1988b. Fibroblast-dependent growth of mouse mast cells *in vitro*: Duplication of mast cell depletion in mutant mice of W/W^v genotype. *J. Cell Physiol.* 134: 78.

Geissler, E.N., M.A. Ryan, and D.E. Housman. 1988. The dominant-white spotting (W) locus of the mouse encodes the c-*kit* proto-oncogene. *Cell* 55: 185.

Giebel, L.B. and R.A. Spritz. 1991. Mutation of the c-*kit* (mast/stem cell growth factor receptor) proto-oncogene in human piebaldism. *Proc. Natl. Acad. Sci.* 88: 8696..

Godin, I., R. Deed, J. Cooke, K. Zsebo, M. Dexter, and C.C. Wylie. 1991. Effects of the *steel* gene product on mouse primordial germ cells in culture. *Nature* 352: 807.

Gruneberg, H. and G.M. Truslove. 1960. Two closely linked genes in the mouse. *Genet. Res.* 1: 69.

Hafen, E., K. Basler, J.-E. Edstroem, and G.M. Rubin. 1987. *Sevenless*, a cell-specific homeotic gene of *Drosophila*, encodes a putative transmembrane receptor with a tyrosine kinase domain. *Science* 236: 55.

Hanks, S.K, A.-M. Quinn, and T. Hunter. 1988. The protein kinase family: Conserved features and deduced phylogeny of the catalytic domains. *Science* 241: 42.

Harrison, D.E. and E.S. Russell. 1972. The response of the W/W^v and Sl/Sl^d anaemic mice to haemopoietic stimuli. *Br. J. Haematol.* 22: 155.

Hayashi, S.I., T. Kunisada, M. Ogawa, K. Yamaguchi, and S.I. Nishikawa. 1991. Exon skipping by mutation of an authentic splice site of c-*kit* gene in W/W mouse. *Nucleic Acids Res.* 19: 1267.

Herskowitz, I. 1987. Functional inactivation of genes by dominant negative mutations. *Nature* 329: 219.

Hertwig, P. 1942. Neue mutationen and koppelungsgruppen bei der hausmaus. *Z. Indukt. Abstamm. Vererbungsl.* 80: 220.

Huang, E., K. Nocka, D.R. Beier, T.-Y. Chu, J. Buck, H.-W. Lahm, D. Wellner, P. Leder, and P. Besmer. 1990. The hematopoietic growth factor KL is encoded at the *Steel* locus and is the ligand of the c-*kit* receptor, the gene product of the W locus. *Cell* 63: 225.

Jarboe, D.L. and T.F. Huff. 1989. The mast cell-committed progenitor. II. W/W^v mutant mice do not make mast cell-committed progenitors and Sl/Sl^d fibroblasts do not support development of normal mast cell-committed

progenitors. *J. Immunol.* **142:** 2418.
Kacser, H. and J.A. Burns. 1981. The molecular basis of dominance. *Genetics* **97:** 639.
Kaneko, Y., J. Takenawa, O. Yoshida, K. Fujita, K. Sugimoto, H. Nakayama, and J. Fujita. 1991. Adhesion of mouse mast cells to fibroblasts: Adverse effects of steel (*sl*) mutation. *J. Cell. Physiol.* **147:** 224.
Kazlauskas, A. and J.A. Cooper. 1989. Autophosphorylation of the PDGF receptor in the kinase insert region regulates interactions with cell proteins. *Cell* **58:** 1121.
Keshet, E., S.D. Lyman, D.W. Williams, D.M. Anderson, N.A. Jenkins, N.G. Copeland, and L.F. Parada. 1991. Embryonic RNA expression patterns of the c-*kit* receptor and its cognate ligand suggest multiple functional roles in mouse development. *EMBO J.* **10:** 2425.
Kitamura, Y. and S. Go. 1979. Decreased production of mast cells in Sl/Sl^d anemic mice. *Blood* **53:** 492.
Kitamura, Y., S. Go, and K. Hatanaka. 1978. Decrease of mast cells in W/W^v mice and their increase by bone marrow transplantation. *Blood* **53:** 447.
Koch, C.A., D. Anderson, M.F. Moran, C. Ellis, and T. Pawson. 1991. SH2 domains: Elements that control interactions of cytoplasmic signaling proteins. *Science* **252:** 668.
Kramer, H., R.L. Cagan, and S.L. Zipursky. 1991. Interaction of *bride of sevenless* membrane-bound ligand and the *sevenless* tyrosine kinase receptor. *Nature* **352:** 207.
Kuroda, H., N. Terada, H. Nakayama, K. Matsumoto, and Y. Kitamura. 1988. Infertility due to growth arrest of ovarian follicles in Sl/Sl^t mice. *Dev. Biol.* **126:** 71.
Kuroda, H,. H. Nakayama, M. Namiki, K. Matsumoto, Y. Nishimune, and Y. Kitamura. 1989. Differentiation of germ cells in seminiferous tubules transplanted to testes of germ cell-deficient mice of W/W^v and Sl/Sl^d genotypes. *J. Cell. Physiol.* **139:** 329.
Lai, C. and G. Lemke. 1991. An extended family of protein-tyrosine kinase genes differentially expressed in the vertebrate nervous system. *Neuron* **6:** 691.
Lyon, M.F. and P.H. Glenister. 1982. A new allele sash (W^{sh})) at the *W*-locus and a spontaneous recessive lethal in mice. *Genet. Res.* **39:** 315.
Lyon, M.F. and A.G. Searle. 1989. *Genetic variants and strains of the laboratory mouse.* Oxford University Press, United Kingdom.
Lyon, M.F., P.H. Glenister, J.F. Loutit, E.P. Evans, and J. Peters. 1984. A presumed deletion covering the *W* and *Ph* loci of the mouse. *Genet. Res.* **44:** 161.
Majumder, S., K. Brown, F.-H. Qiu, and P. Besmer. 1988. c-*kit* protein, a transmembrane kinase: Identification in tissues and characterization. *Mol. Cell. Biol.* **8:** 4896.
Manova, K. and R.F. Bachvarova. 1991. Expression of c-*kit* encoded at the *W* locus of mice in developing embryonic germ cells and presumptive melanoblasts. *Dev. Biol.* **146:** 312.
Manova, K., K. Nocka, P. Besmer, and R.F. Bachvarova. 1990. Gonadal expression of c-*kit* encoded at the *W* locus of the mouse. *Development* **110:** 1057.
Marks, S.C., Jr. 1984. Congenital osteopetrotic mutations as probes of the origin, structure and function of osteoclasts. *Clin. Orthop. Relat. Res.* **189:** 239.
Martin, F.H., S.V. Suggs, K.E. Langley, H.S. Lu, J. Ting, K.H. Okino, C.F. Morris,

I.K. McNiece, F.W. Jacobsen, E.A. Mendiaz, N.C. Birkett, K.A. Smith, M.J. Johnson, V.P. Parker, J.C. Flores, A.C. Patel, E.F. Fisher, H.O. Erjavec, C.J. Herrera, J. Wypych, R.K. Sachdev, J.A. Pope, I. Leslie, D. Wen, C.-H. Lin, R.L Cupples, and K.M. Zsebo. 1990. Primary structure and functional expression of rat and human stem cell factor DNAs. *Cell* **63**: 203.

Matsui, Y., K.M. Zsebo, and B.L.M. Hogan. 1990. Embryonic expression of a hematopoietic growth factor encoded by the *Sl* locus and the ligand for c-*kit*. *Nature* **347**: 667.

Matthews, W., C.T. Jordan, G.W. Wiegand, D. Pardoll, and J.R. Lemischka. 1991. A receptor tyrosine kinase specific to hematopoietic stem and progenitor cell-enriched populations. *Cell* **65**: 1143.

Mayer, T.C. 1970. A comparison of pigment cell development in albino, Steel and dominant spotting mutant mouse embryos. *Dev. Biol.* **23**: 297.

Mayer, T.C. and M.C. Green. 1968. An experimental analysis of the pigment defect caused by mutations at the W and *Sl* loci in mice. *Dev. Biol.* **18**: 62.

McCoshen, J.A. and D.J. McCallion. 1975. A study of the primordial germ cells during migratory phase in steel mutant mice. *Experientia* **31**: 589.

McCulloch, E.A., L. Siminovitch, and J.E. Till. 1964. Spleen-colony formation in anemic mice of genotype W/W^v. *Science* **144**: 844.

McCulloch, E.A., L. Siminovitch, J.E. Till, E.S. Russell, and S.E. Bernstein. 1965. The cellular basis of the genetically determined hemopoietic defect in anemic mice of genotype Sl/Sl^d. *Blood* **26**: 399.

McNiece, I.K, K.E Langley, and K.M. Zsebo. 1991a. Recombinant human stem cell factor synergises with GM-CSF, G-CSF, IL-3 and Epo to stimulate human progenitor cells of the myeloid and erythroid lineages. *Exp. Hematol.* **19**: 226.

―――. 1991b. The role of recombinant stem cell factor in early B cell development. *J. Immunol.* **146**: 3785.

Metcalf, D. and N.A. Nicola. 1991. Direct proliferation of actions of stem cell factor on murine bone marrow cells *in vitro*: Effects of combination with colony-stimulating factors. *Proc. Natl. Acad. Sci.* **88**: 6239.

Migliaccio, G., A.R. Migliaccio, J. Valinsky, K. Langley, K. Zsebo, J.W.M. Visser, and J.W. Adamson. 1991. Stem cell factor induces proliferation and differentiation of highly enriched murine hematopoietic cells. *Proc. Natl. Acad. Sci.* **88**: 7420.

Mintz, B. and E.S. Russell. 1957. Gene-induced embryological modifications of primordial germ cells in the mouse. *J. Exp. Zool.* **134**: 207.

Moran, M.F., C.A Koch, I. Sadowski, and T. Pawson. 1988. Mutational analysis of a phosphotransfer motif essential for v-*fps* tyrosine kinase activity. *Oncogene* **3**: 665.

Motro, B., D. van der Kooy, J. Rossant, A. Reith, and A. Bernstein. 1991. Contiguous patterns of c-*kit* and *steel* expression: Analysis of mutations at the W and *Sl* loci. *Development* (in press).

Nakayama, H., H. Kuroda, H. Onoue, J. Fujita, Y. Nishimune, K. Matsumoto, T. Nagano, F. Suzuki, and Y. Kitamura. 1988. Studies of Sl/Sl^d ↔ +/+ mouse aggregation chimaeras II. Effect of the steel locus on spermatogenesis. *Development* **102**: 117.

Nishikawa, S., M. Ogawa, M. Kusakabe, S.-I. Hayashi, T. Kunisada, T. Era, T. Sakakura, and S.-I. Nishikawa. 1990. *In utero* manipulation of coat color formation by a monoclonal anti-c-*kit* antibody: Two distinct waves of c-*kit*

dependency during melanocyte development. *EMBO J.* **8:** 2111.
Nishimune, Y., T. Hanegi, M. Maekawa, and Y. Kitamura. 1984. In vitro demonstration of deleterious effect of steel mutation on spermatogenesis in mice. *J. Cell. Physiol.* **118:** 19.
Niwa, Y., T. Kasugai, K. Ohno, M. Morimoto, M. Yamazaki, K. Dohmae, Y. Nishimune, K. Kondo, and Y. Kitamura. 1991. Anemia and mast cell depletion in mutant rats that are homozygous at "White Spotting (Ws)" locus. *Blood* **78:** 1936.
Nocka, K., J. Buck, E. Levi, and P. Besmer. 1990a. Candidate ligand for the c-*kit* transmembrane kinase receptor: KL, a fibroblast derived growth factor stimulates mast cells and erythroid progenitors. *EMBO J.* **9:** 3287.
Nocka, K., S. Majumder, B. Chabot, P. Ray, M. Cervone, A. Bernstein, and P. Besmer. 1989. Expression of the c-*kit* proto-oncogene in known cellular targets of *W* mutations in normal and *W* mutant mice: Evidence for an impaired c-*kit* kinase in mutant mice. *Genes Dev.* **3:** 816.
Nocka, K., J.C. Tan, E. Chiu, T.Y. Chu, P. Ray, P. Traktman, and P. Besmer. 1990b. Molecular bases of dominant negative and loss of function mutations at the murine c-*kit*/white spotting locus: W^{37}, W^v, W^{41} and W. *EMBO J.* **9:** 1805.
Ogawa, M., Y. Matsuzaki, S. Nishikawa, S.I. Hayashi, T. Kunisada, T. Sudo, T. Kina, H. Nakauchi, and S.I. Nishikawa. 1991. Expression and function of c-*kit* in hemopoietic progenitor cells. *J. Exp. Med.* **174:** (in press).
Olivieri, N.F., T. Grunberger, Y. Ben-David, J. Ng, D.E. Williams, S. Lyman, D.M. Anderson, A.A. Axelrad, P. Correa, A. Bernstein, and M.N. Freedman. 1991. Diamond-Blackfan anemia: Heterogenous response of hematopoietic progenitor cells *in vitro* to the protein product of the *steel* locus. *Blood* (in press).
Orr-Urtreger, A., A. Avivi, Y. Zimmer, D. Givol, Y. Yarden, and P. Lonai. 1990. Developmental expression of c-*kit*, a proto-oncogene encoded by the *W* locus. *Development* **109:** 911.
Papayannopoulou, T., M. Brice, V.C. Broudy, and K.M. Zsebo. 1991. Isolation of c-*Kit* receptor-expressing cells from bone marrow, peripheral blood, and fetal liver: Functional properties and composite antigenic profile. *Blood* **78:** 1403.
Pawson, T. and A. Bernstein. 1990. Receptor tyrosine kinases: Genetic evidence for their role in *Drosophila* and mouse development. *Trends Genet.* **6:** 359.
Qiu, F., P. Ray, X. Brown, P.E. Barker, S. Jhanwar, F.H. Ruddle, and P. Besmer. 1988. Primary structure of c-*kit*: Relationship with the CSF-1/PDGF receptor kinase family—oncogenic activation of v-*kit* involves deletion of extracellular domain and C terminus. *EMBO J.* **7:** 1003.
Reith, A.D., R. Rottapel, E. Giddens, C. Brady, L. Forrester, and A. Bernstein. 1990. *W* mutant mice with mild or severe developmental defects contain distinct point mutations in the kinase domain of the c-*kit* transmembrane receptor. *Genes Dev.* **4:** 390.
Reith, A.D., C. Ellis, S.D. Lyman, D.M. Anderson, D.E. Williams, A. Bernstein, and T. Pawson. 1991. Signal transduction by normal isoforms and *W* mutant variants of the Kit receptor tyrosine kinase. *EMBO J.* **10:** 2451.
Roberts, W.M., A.T. Look, M.F. Roussel, and C.J. Sherr. 1988. Tandem linkage of human CSF-1 receptor (c-*fms*) and PDGF receptor genes. *Cell* **55:** 655
Rosnet, O., M.-G. Mattei, S. Marchetto, and D. Birnbaum. 1991. Isolation and

chromosomal localization of a novel FMS-like tyrosine kinase gene. *Genomics* **9:** 380.

Rossant, J.R. and A. Joyner. 1989. Towards a molecular-genetic analysis of mammalian development. *Trends Genet.* **5:** 277.

Rottapel, R., M. Reedijk, D.E. Williams, S.E. Lyman, D.M. Andersen, T. Pawson, and A. Bernstein. 1991. The *Steel/W* signal transduction pathway: Kit autophosphorylation and its association with a unique subset of cytoplasmic signaling proteins is induced by the Steel factor. *Mol. Cell. Biol.* **11:** 3043.

Rubin, G.M. 1989. Development of the *Drosophila* retina: Inductive events studied at single cell resolution. *Cell* **57:** 519.

Russell, E.S. 1979. Hereditary anemias of the mouse: A review for geneticists. *Adv. Genet.* **20:** 357.

Sarvella, P.A. and L.B. Russell. 1956. Steel, a new dominant gene in the house mouse. *J. Hered.* **47:** 123.

Schlager, G. and M.M. Dickie. 1971. Natural mutation rates in the house mouse: Estimates for five specific loci and dominant mutations. *Mutat. Res.* **11:** 89.

Searle, A.G. and G.M. Truslove. 1970. A gene triplet in the mouse. *Genet. Res.* **15:** 227.

Sherr, C.J. 1990. Colony-stimulating factor-1 receptor. *Blood* **75:** 1.

Shibuya, M., S. Yamaguchi, A. Yamane, T. Ikeda, A. Tojo, H. Matsushime, and M. Sato. 1990. Nucleotide sequence and expression of a novel human receptor-type tyrosine kinase gene (*flt*) closely related to the *fms* family. *Oncogene* **5:** 519.

Silvers, W.K. 1979. White-spotting, patch and rump-white. In *The coat colors of mice: A model for gene action and interaction*, p. 206. Springer-Verlag, New York.

Smith, E.A, M.F. Seldin, L. Martinez, M.L Watson, G.G. Choudhury, P.A Lalley, J. Pierce, S. Aaaronson, J. Barker, S.L. Naylor, and A.Y. Sakaguchi. 1991. Mouse platelet-derived growth factor receptor α gene is deleted in W[19H] and patch mutations on chromosome 5. *Proc. Natl. Acad. Sci.* **88:** 4811.

Stephenson, D.A., M. Mercola, C. Wang, C.D. Stiles, D.F. Bowen-Pope, and V.M. Chapman. 1991. Platelet-derived growth factor receptors α-subunit gene (*pdgfra*) is deleted in the mouse patch (*Ph*) mutation. *Proc. Natl. Acad. Sci.* **88:** 6.

Tan, J.C., K. Nocka, P. Ray, P. Traktman, and P. Besmer. 1990. The dominant W[42] spotting phenotype results from a missense mutation in the c-*kit* receptor. *Science* **247:** 209.

Tsujimura, T., S. Hirota, S. Nomura, Y. Niwa, M. Yamazaki, T. Tono, E. Morii, Y.-M. Kim, K. Kondo, Y. Nishimune, and Y. Kitamura. 1991. Characterization of Ws mutant allele of rats: A 12 base deletion in tyrosine kinase domain of c-*kit* gene. *Blood* **78:** 1942..

Ullrich, A. and J. Schlessinger. 1990. Signal transduction by receptors with tyrosine kinase activity. *Cell* **61:** 203.

Walker, D.G. 1975. Spleen cells transmit osteopetrosis in mice. *Science* **190:** 785.

Williams, D.E., J. Eisenman, A. Baird, C. Rauch, V.K. Ness, C.J. March, L.S. Park, U. Martin, D.Y. Mochizuki, H.S. Boswell, G.S. Burgess, D. Cosman, and S.D. Lyman. 1990. Identification of a ligand for the c-*kit* proto-oncogene. *Cell* **63:** 167.

Yarden, Y. and A. Ullrich. 1988. Growth factor receptor tyrosine kinases. *Annu. Rev. Biochem.* **57:** 443.

Yarden, Y., W.J. Kuang, T. Yang-Feng, L. Coussens, S. Munemitsu, T.J. Dull, E. Chen, J. Schlessinger, U. Francke, and A. Ullrich. 1987. Human proto-oncogene c-*kit*: A new cell surface receptor tyrosine kinase for an unidentified ligand. *EMBO J.* **6:** 3341.

Yoshida, H., S.-I. Hayashi, T. Kunisada, M. Ogawa, S. Nishikawa, H. Okamura, T. Sudo, L.D. Shultz, and S.-I. Nishikawa. 1990. The murine mutation osteopetrosis is in the coding region of the macrophage colony stimulating factor gene. *Nature* **345:** 442.

Yoshinaga, K., S. Nishikawa, M. Ogawa, S.-I. Hayashi, T. Kunisada, T. Fujimoto, and S.-I. Nishikawa. 1991. Role of c-*kit* in mouse spermatogenesis: Identification of spermatogonia as a specific site of c-*kit* expression and function. *Development* (in press).

Yu, J.-C., M.A Heidaran, J.H. Pierce, J.S. Gutkind, D. Lombardi, M. Ruggiero, and S.A. Aaronson. 1991. Tyrosine mutations within the platelet-derived growth factor receptor kinase insert domain abrogate receptor-associated phosphatidylinositol-3 kinase activity without affecting mitogenic or chemotactic signal transduction. *Mol. Cell. Biol.* **11:** 3780.

Zsebo, K.M., J. Wypych, I.K. McNeice, H.S. Lu, K.A. Smith, S.B. Karkare, R.K. Sachdev, V.N. Yuschenkoff, N.C. Birkett, L.R. Williams, V.N. Satyagel, W. Tung, R.A Bosselman, E.A. Mendiaz, and K.E. Langley. 1990a. Identification, purification and biological characterization of hemopoietic stem cell factor from Buffalo rat liver-conditioned medium. *Cell* **63:** 195.

Zsebo, K.M., D.A. Williams, E.N. Geissler, V.C. Broudy, F.H. Martin, H.L. Atkins, R.-Y. Hsu, N.C. Birkett, K.H. Okino, D.C. Murdock, F.W. Jacobsen, K.E. Langley, K.A. Smith, T. Takeishi, B.M. Cattanach, S.J. Galli, and S.V. Suggs. 1990b. Stem cell factor is encoded at the *Sl* locus of the mouse and is the ligand for the c-*kit* tyrosine kinase receptor. *Cell* **63:** 213.

Molecular Genetics of Wilms' Tumor

Jerry Pelletier,[1,2] David Munroe,[1] and David Housman[1]

[1]Massachusetts Institute of Technology
Center for Cancer Research
Cambridge, Massachusetts 02139

[2]McGill Cancer Center
McGill University
Montreal, Quebec, Canada, H3G 1Y6

The short arm of chromosome 11 is one of the most intensively studied regions of the human genome. Much of the impetus to this study has been the identification within this chromosomal region of a tumor suppressor gene for a prominent form of childhood cancer—Wilms' tumor.

Herein we review:

❑ clinical findings in both Wilms' tumor and associated syndromes

❑ genetic analysis that led to the identification of the Wilm's tumor suppressor gene

❑ physical mapping of the region surrounding the gene

❑ localization of additional genes of clinical significance in the region

❑ characterization of the 11p13 Wilms' tumor gene

INTRODUCTION

Wilms' tumor (WT) or nephroblastoma, the most common primary renal tumor in children, accounts for about 10% of pediatric malignant tumors (Aron 1974). Although previous workers had described the same tumor, Max Wilms first recognized nephroblastoma as a clinical entity, and to this date, the eponym "Wilms' tumor" persists (Wilms 1899). The occurrence of this disease is remarkably constant among diverse population groups differing in environment and life style and exhibiting highly variable incidence rates for most other cancers (Innis 1972). Regardless of race, sex, or country of birth, about 1 in 10,000 children under 15 years of age may be expected to develop WT, with the peak incidence occurring between the ages of 3 and 4 years (Pendergrass 1976; Matsunaga 1981). Metastasis of WT spreads primarily to the lungs and liver and rarely to bone and brain (D'Angio et al. 1989).

In 1938, a survey of 383 cases recorded a cure rate of 5.7% (McNeill and Chilko 1938). Ladd (1938) showed that with improved surgical techniques, the survival rate increased to 20%. In 1950, Gross and Neuhauser (1950) indicated that postoperative radiotherapy increased the cure rate to 50%. With the addition of chemotherapy to the antitumor regimen, combined with more aggressive treatment of the metastatic disease, survival rates currently approach 90% (D'Angio et al. 1989). The treatment of WT remains one of the clearest examples of success in pediatric oncology.

WILMS' TUMOR-ASSOCIATED SYNDROMES

Wilms' tumors are often composed of three elements—blastemal, epithelial, and stromal—the ratio of which varies among tumors. Nodular renal blastema (persistent nephrogenic rests) is thought to be a precursor lesion of WT, having been identified in as many as 40% of patients with WT (Bove et al. 1969; Bennington and Beckwith 1975). WTs are thought to arise from the metanephric blastema of the developing kidney during intrauterine and/or early extrauterine life. In situ nephroblastomas have been found in 0.4 % of autopsied infants less than 4 months old, suggesting that WT may be more prevalent than indicated by the incidence rate, with many undergoing spontaneous regression (Shanklin and Sotelo-Avila 1969).

The WAGR syndrome

Most cases of WT are unilateral, although some 5–10% are bilateral. The age of onset is usually much earlier in bilateral than in unilateral cases (Matsunaga 1981). On the basis of a study of more than 500 cases of WT, Knudson and Strong (1972) postulated a "two-step" mutational model to account for the incidence of sporadic and bilateral forms (see Fig. 1). According to this hypothesis, the rate-limiting events for development of sporadic tumors are the occurrence of two independent postzygotic mutational events within the same cell. The majority of bilateral tumors are presumed to arise because affected individuals carry a germ line mutation, and only one additional lesion in the target cell is required for tumorigenesis. Hereditary cases of WT display both a high penetrance and an increased incidence of bilateral tumors, attesting to the frequency of the second somatic event.

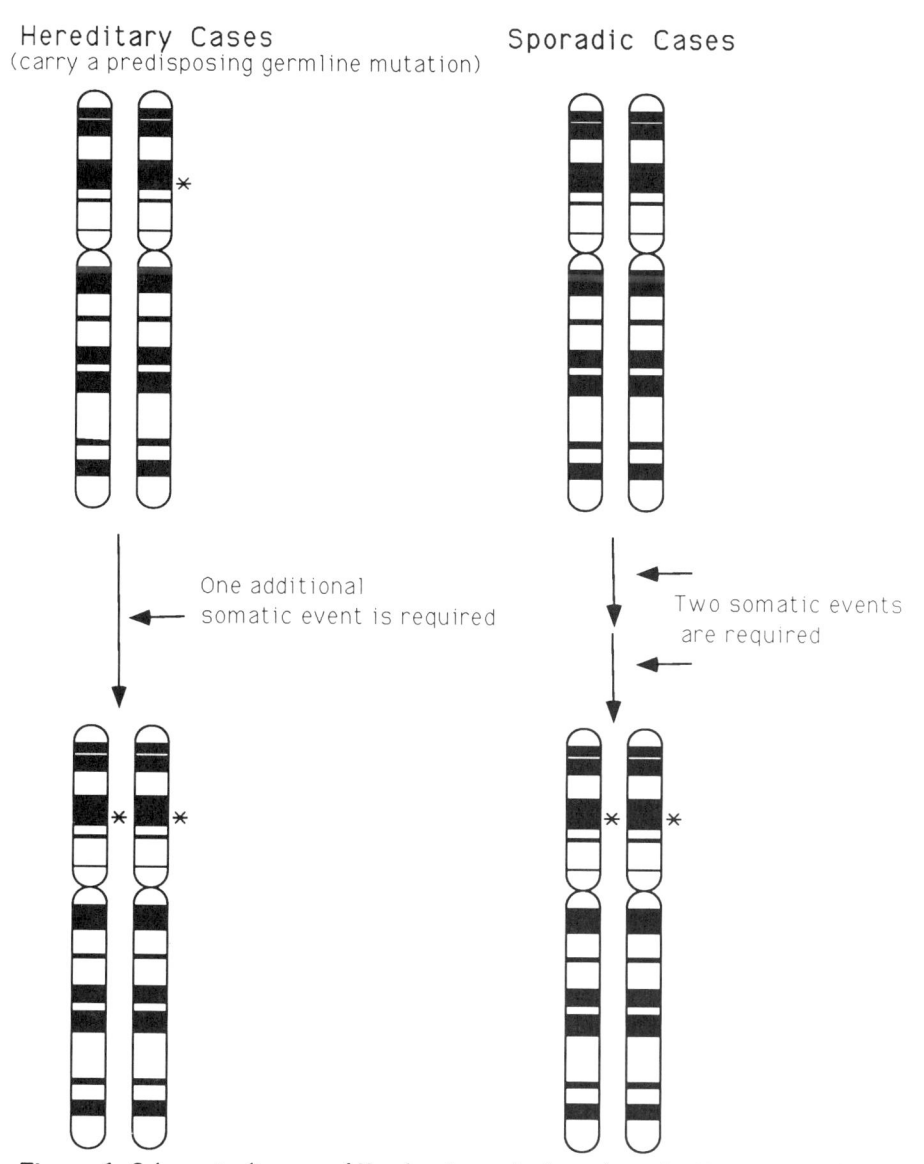

Hereditary Cases
(carry a predisposing germline mutation)

Sporadic Cases

← One additional somatic event is required

← Two somatic events are required →

Figure 1 Schematic diagram of Knudson's two-hit hypothesis for WT. See text for explanation.

In 1964, Miller, Fraumeni, and Manning noted a crucial statistical association between the incidence of WT and another rare disease, bilateral aniridia (Miller et al. 1964). The frequency of occurrence of aniridia in the general population is estimated as 1 in 50,000–100,000 (Shaw et al. 1960; Francois et al. 1977). Miller et al. (1964) found that the rate of occurrence of aniridia among patients with WT was 1 in 73, about 1000-fold higher than would occur by chance. In addition, aniridic children tend to be younger when their tumors are first diagnosed (average, 1.8 years) than other children with WT without concurrent aniridia (average, 3.6 years) (Miller et al. 1964; Fraumeni and Glass 1968). WT occurs in 33% of patients with sporadic aniridia (Broduer 1984). Patients with aniridia who develop WT, have a greater incidence of bilateral involvement than do WT patients in general (Miller et al. 1964; Bond 1975). These observations suggest that the WT and aniridia genes are in close physical proximity.

As a result of further clinical investigation, the aniridia-WT association was expanded to include genitourinary (GU) anomalies and mental retardation. GU anomalies involve the kidney, collecting system, and external genitalia, and include renal hypoplasia, unilateral renal agenesis, horseshoe kidneys, uretral atresia, bifid ureters, hypospadias (misplaced external penile urinary orifice), cryptorchidism (undescended testis), and ambiguous genitalia (Miller et al., 1964; Pendergrass 1976; Breslow and Beckwith 1982). The frequency of occurrence of GU anomalies among the general population is quite high (0.3% for hypospadias and 1.5% for cryptorchidism) (Horton and Devine 1972; Breslow and Beckwith 1982). Their incidence in association with sporadic WT is approximately threefold higher (Miller et al. 1964; Breslow and Beckwith 1982). However, the incidence of congenital abnormalities in children with *bilateral* WT (predicted by the Knudson hypothesis to carry a germ line mutation within the WT gene) is at least ten times higher than that for patients with unilateral disease (Bond 1975; Breslow and Beckwith 1982). GU defects are more often associated with bilateral WT than is aniridia (Pendergrass 1976), suggesting a tighter linkage between the GU and WT genes than between the aniridia and WT genes.

Retardation, a fourth phenotype often observed in patients with WT and aniridia, includes mental retardation, delayed somatic growth, and cognitive and motor impairment. The association of these four symptom groups has led to the acronym WAGR syndrome for *W*ilms' tumor, *a*niridia, *g*enitourinary malformations, and *r*etardation. An additional, potentially important, auxiliary phenotype to the WAGR syndrome is a set of cardiac malformations that have been described for a number of WAGR patients (Miller et al. 1964; Gilgenkrantz et al. 1982; van Heyningen et al. 1985; Glaser 1988; Stiller et al. 1987). The frequency of septal defects among aniridic children with WT was found to be approximately 16%, compared to an overall rate of 0.35% for all septal defects (Stiller et al. 1987).

Patients showing karyotypic rearrangements of 11p An explanation for the association of the clinical phenotypes of the WAGR syndrome began to emerge in the late 1970s as improved cytogenetic techniques revealed constitutional deletions of chromosome 11 in these patients. Initial studies by Riccardi et al. (1978) and Francke et al. (1979) defined a critical role for band p13 of chromosome 11 in the WAGR syndrome. In addition, the description of patients with apparently normal karyotypes, yet who exhibited WT in combination with either aniridia or GU anomalies, suggested that the genes responsible for these phenotypes are sufficiently close to be affected by a presumably submicroscopic deletion (Ferrell and Riccardi 1981; Niikawa et al. 1982; Riccardi et al. 1982; Nakagome et al. 1984). The association between genetic transmission of *del*(11p13) and development of the WAGR phenotypes underscored the involvement of this chromosomal region in hereditary cases of WT (Yunis and Ramsay 1980; Nakagome et al. 1984; Lavedan et al. 1989). The association between WT and loss of genetic information at 11p13 led to the proposed existence of a recessive oncogene involved in the etiology of WT located in this chromosomal region.

Deletion of 11p13 does not lead to WT with 100% penetrance. This is illustrated, for example, by a case of monozygous twins who both had *del*(11p) and aniridia, yet only one child developed bilateral WT (Cotlier et al. 1978). When WT is found in a familial setting, its inheritance follows an autosomal dominant pattern with an estimated penetrance of approximately 0.4 (Matsunaga 1981). In hereditary cases, the proportion of bilateral involvement is higher (~20%) than in the sporadic form (Matsunaga 1981). The age of onset of hereditary unilateral disease is much lower (~31.9 months) than for sporadic cases (~46.4 months). These data are consistent with hereditary cases carrying a predisposing germ line mutation within the WT gene, as predicted by the Knudson hypothesis.

A second form of karyotypic rearrangement disrupting 11p13 has provided genetic evidence that the gene responsible for the aniridia phenotype is distinct from the WT gene. Only recently has autosomal dominant aniridia been shown to be linked to chromosome 11p13 markers (Mannens et al. 1989). Two families have been described with translocations disrupting 11p13 in which the presence of the translocated chromosome correlates directly with the occurrence of aniridia (Simola et al. 1983; Moore et al. 1986). Constitutional inactivation of one allele of the aniridia gene seems sufficient to confer the aniridia phenotype. The penetrance of familial aniridia is essentially complete. No WT was observed in family members carrying the *der*(11) chromosome, suggesting that it disrupted the aniridia locus without affecting the WT gene.

Karyotypic rearrangements found in WT Cytogenetic analysis of WT has provided valuable information on the chromosomal changes that occur during tumorigenesis. We have reviewed 136 cases of karyotype analysis of WT reported in the literature (Kaneko et al. 1981, 1983, 1991; Osada et al. 1981; Slater and de Kraker 1982; Ferrell et al. 1983; Gardner et al. 1983; Kondo et al. 1984; Douglass et al. 1985; Slater et al. 1985; Kovacs et al. 1987; Kumar et al. 1987; Wolman et al. 1987; Solis et al. 1988; McDowell 1989; Sherwood et al. 1989; Diaz de Bustamante et al. 1990; Wang-Wuu et al. 1990). The most common chromosome structural abnormalities observed in WT occur on chromosomes 1, 11p, 7, and 16. No particular regions of chromosomes 1, 7, and 16 are consistently involved in rearrangements, arguing that their involvement is secondary, rather than primary. However, the specific involvement of 11p13 in tumorigenesis is substantiated by the finding that in approximately 20% of all tumors analyzed, 11p13 is hemizygously deleted or involved in translocations.

Abnormal chromosome numbers in WT were first reported by Cox (1966). Wang-Wuu et al. (1990) have reported a karyotype study of 31 WTs in which chromosome 12 was trisomic in 13 of 25 tumors. They suggest that possibly a gene-dosage effect of c-Ki-*ras* (which is located on chromosome 12) could contribute to the neoplastic phenotype. In addition, they reported an occurrence of 28% for trisomy 18 in their series of tumors. This numerical change is of particular interest since patients with constitutional trisomy 18, at postmortem, commonly have nephroblastoma in situ (Bove et al. 1969) and, in one case, a WT (Geiser and Shindler 1969). Characteristics of trisomy 18 children are variable and include severe mental retardation, heart defects, odd-shaped heads, and stunted growth. Survival beyond 1 year is rare.

The significance of apparently secondary chromosome changes in the progression of WT is not known. It is possible that some of these cytogenetic abnormalities reflect a molecular lesion, outside 11p, required for tumorigenesis (see below).

The Beckwith-Wiedemann syndrome

A second WT-associated syndrome is Beckwith-Wiedemann syndrome (BWS). This syndrome is characterized by exomphalos (umbilical hernia), macroglossia (enlargement of the tongue), neonatal hypoglycemia, gigantism, and increased risk for development of specific childhood cancers (Beckwith 1963; Wiedemann 1964). The characteristic features of BWS may be quite variable with no strictly defined minimal criteria identifying all patients. This is particularly exemplified by the noted association of hemihypertrophy and WT (Miller et al. 1964). These individuals are also predisposed to developing adrenal neoplasms and hepatoblastomas. Many investigators think that hemihypertrophy represents the mild end of the BWS spectrum. The prevalence of BWS has been calculated to be approximately 7 per 100,000 births (Higurashi et al. 1980; Pettenati et al. 1986). Sotelo-Avila et al. (1980) noted that of 200 reported cases of BWS, 10 had developed nephroblastoma, 4 had developed adrenocortical carcinoma, and 1 had developed hepatoblastoma. Wiedemann (1983) found that of 388 BWS infants, 29 developed a total of 32 tumors, giving a risk estimate of 7.5%. Of these tumors, 14 were WT, 5 were adrenal carcinomas, and 2 were hepatoblastomas.

Most cases of BWS are sporadic, although several familial cases have been reported. Genetic linkage analysis of BWS families has revealed that the BWS lesion is linked to 11p15.5 (Koufos et al. 1989; Joy Ping et al. 1989). BWS exhibits a dominant pattern of inheritance with variable expressivity (Niikawa et al. 1986).

A specific chromosomal abnormality relating to BWS was first documented by Waziri et al. (1983), who reported two unrelated patients with 11p duplications. Subsequently, several other groups reported similar findings (Turleau et al. 1984; Journel et al. 1985; Wales et al. 1986; Henry et al. 1989a). The majority of sporadic BWS cases, however, do not show cytogenetically visible chromosomal rearrangements. Analyses of gene dosage with molecular probes from chromosome 11 in children with BWS have failed to reveal small duplications of the genes for c-Ha-*ras* 1 (*HRAS1*), insulin (*INS*), and insulin-like growth factor II (*IGF-II*) (which are located in 11p15.5, see Fig. 2) (Henry et al. 1988; Schofield et al. 1989). No chromosomal abnormalities have been observed in familial cases of BWS (Pettenati et al. 1986).

Another interesting aspect of BWS is the possibility that genomic imprinting may play a role in the expression of its clinical features. The expressivity of BWS has been noted to depend on inheritance from either father or mother (Lubinsky et al. 1974; Niikawa et al. 1986). If 11p15.5 is imprinted in humans, this raises the interesting possibility that this epigenetic phenomenon may affect expressivity of BWS (for review, see Hall 1990). A survey of cases of 11p15 duplications in BWS concluded that when the parental origin of the duplicated material could be ascertained, it is always of paternal origin (Brown et al. 1990). These findings suggest that the maternal chromosome 11 has some role in suppression (at 11p15.5) not compensated for by the paternal chromosome 11 (see below).

An interesting candidate gene for at least some features of BWS is *IGF-II*. The gene encoding this growth factor maps to 11p15.5 immediately adjacent to the insulin gene (Bell et al. 1985). IGF-I and IGF-II are polypeptide hormones related to insulin by structure and function. The insulin-like effects of IGF-I and -II are similiar in vitro (Rinderknecht and Humbel 1976). In vivo, IGF-I mimics the effects of growth hormone and is therefore considered to be a major somatomedin in humans (Schoenle et al. 1982). The main function of IGF-II in humans is not clear. IGF-II is synthesized almost ubiquitously during the first trimester, with the major sites of mRNA synthesis being the liver, kidney, and adrenal glands (Scott et al. 1985). These are the same organs in which BWS patients are at risk for developing tumors. Recently, a struc-

142 J. PELLETIER, D. MUNROE, AND D. HOUSMAN

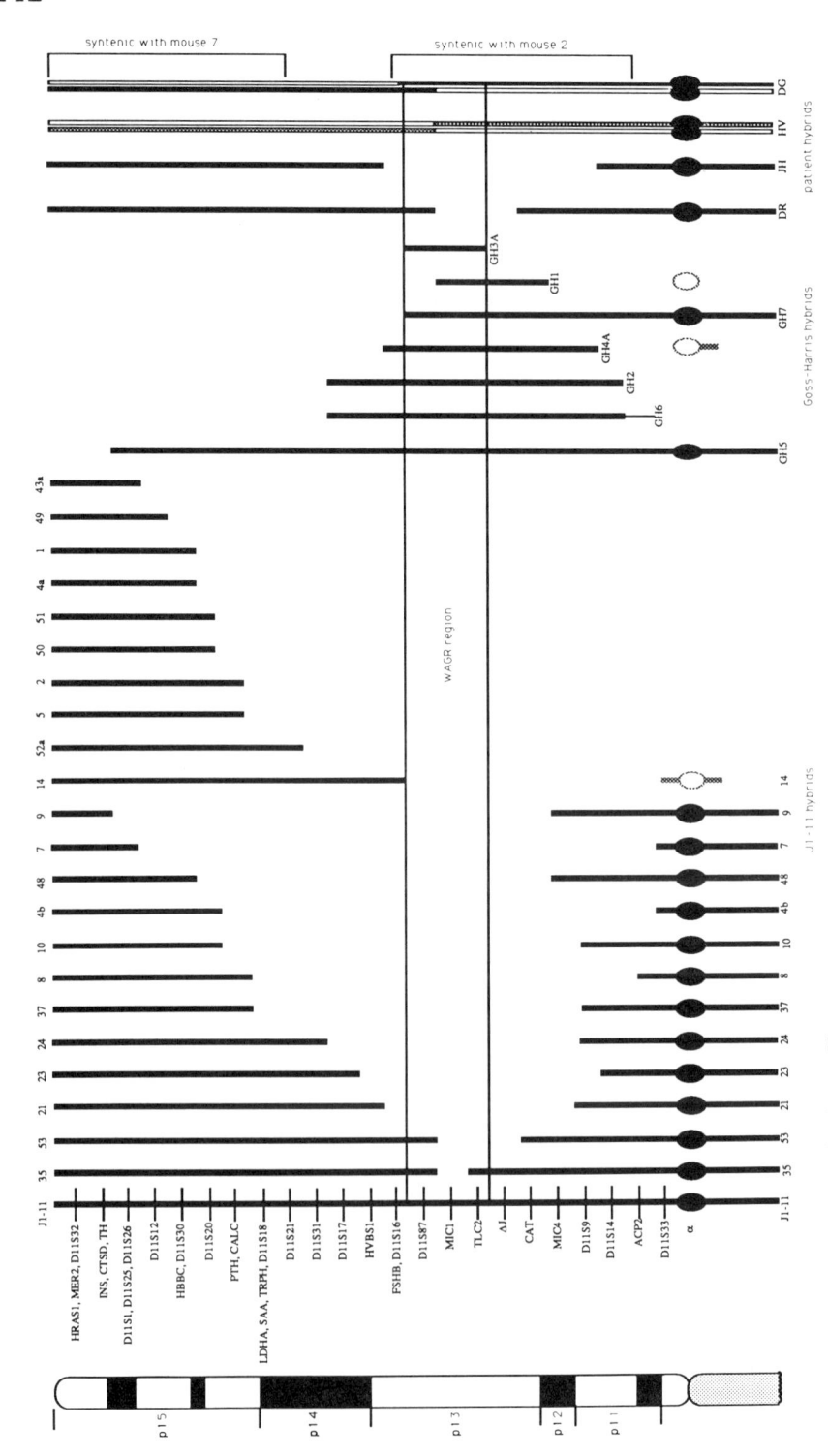

Figure 2 (*See facing page for legend.*)

tural alteration of the *IGF-II* gene in a WT was identified (Irminger et al. 1989), although the specific region of the gene involved was not mapped. Several reports have shown that expression of *IGF-II* mRNA in WT is enhanced to levels comparable to that of fetal tissue. Reeve et al. (1985) and Scott et al. (1985) have shown that in 14 of 16 WTs, *IGF-II* mRNA was increased 10- to 100-fold, whereas the levels of *IGF-I* were normal. It should be noted that in one case of a sporadic WT, *IGF-II* mRNA levels were increased 30-fold compared to normal kidney, yet protein levels were deemed unchanged as determined by an IGF-II bioassy (Haselbacher et al. 1987).

It has recently been demonstrated that the maternal gene for IGF-II is imprinted in mouse embryos (DeChiara et al. 1991). Neonates that inherited a disrupted *IGF-II* allele from their father have a small body size, whereas neonates that inherit a maternally derived disrupted *IGF-II* allele are of normal size. The differences in phenotype are a result of differences in expression of the parental alleles (possibly due to genomic imprinting). Thus, many features of *IGF-II* make it an interesting BWS candidate gene; however, as noted above, analysis of gene dosage has falled to demonstrate *IGF-II* duplication in BWS patients. The potential role of *IGF-II* in either Wilms' tumorigenesis or BWS remains undefined.

Denys-Drash syndrome

The first report of a WT occurring in a hermaphroditic child appeared in 1912 (Raubitschek 1912). Stump and Garrett (1954) reported a case of a male pseudohermaphrodite with bilateral Wilms' tumor (Stump and Garrett 1954). This patient subsequently developed progressive renal failure. Denys et al. (1967) described an association between pseudohermaphroditism, WT, and renal failure in a child 15 months old. Drash et al. (1970) added to this initial finding by reporting on two unrelated children evaluated for sexual ambiguity. These children developed WT and eventually died as a result of progressive renal failure. The histological changes that occur in the kidney are characterized by glomerular capillary damage with thickening of the endothelial membrane (Drash et

Figure 2 J1 series deletion map of the short arm of human chromosome 11. The 11p contents of J1 derivative cell lines are shown by solid bars. Goss-Harris derivatives of J1-11 are included in the center of the panel of somatic cell hybrids. 35 genetic markers tested on these cell lines are shown to the left. An ideogram of the short arm of chromosome 11 is presented to the left as a reference. The WAGR complex has been located in the center of band 11p13 by the smallest overlap of WAGR patients' deletions (Francke et al. 1979). On the left, reciprocal translocations from two patients with aniridia, but no WT are shown with their 11p breakpoints; open bars represent chromosome 11 segments. These two patients have been described by Glaser et al. (1986). Brackets on the far right frame areas of mouse-human synteny.

al. 1970). The nephropathy progresses rapidly and is unresponsive to steroids, and eventually the patient requires dialysis. The association between nephroblastoma, pseudohermaphroditism, and nephropathy is commonly referred to as Drash syndrome. This syndrome should perhaps be referred to as the Denys-Drash syndrome, as suggested by Garfunkel (1985).

The Wilms' tumors in patients with Drash syndrome may be unilateral, unilateral but multinodular, or bilateral (Denys et al. 1967; Drash et al. 1970; Spear et al. 1971; Barakat et al. 1974; McCoy et al. 1983; Habib et al. 1985; Gallo and Chemes 1987; Jensen et al. 1989; Jadresic et al. 1990; Tank and Melvin 1990). Cases have been reported of infants with gonadal abnormalities and kidney disease, but without WT (Barakat et al. 1974; Habib et al. 1985; Jensen et al. 1989; Jadresic et al. 1990). The majority of these cases, however, died at a very young age or had renal transplants. In several instances, pathological examination of the removed kidneys revealed small foci of nephroblastoma (Jensen et al. 1989).

The genital abnormalities seen in Denys-Drash syndrome range from clitoromegaly in an otherwise phenotypic female patient (Drash et al. 1970) to severe sexual ambiguity. Internal Müllerian duct structures are sometimes found (Goldman et al. 1981; Manivel et al. 1987), and in several cases, grossly normal external and internal female genitalia were present, but the gonads were found to contain only testicular tissue (Drash et al. 1970; Jensen et al. 1989). Thus, a wide range of reproductive tract abnormalities are observed.

In males, sexual differentiation is mediated through the secretion of testosterone and Müllerian-inhibiting substance by the embryonic testis. In target tissues, testosterone is converted to dihydrotestosterone by the enzyme 5-α reductase. Testosterone elicits development of the male (Wolffian) genital duct system, whereas the external genitalia form under the influence of dihydrotestosterone. Deficient or defective binding of dihydrotestosterone to the target cell or deficient or abnormal 5-α reductase activity can result in male pseudohermaphroditism (Wilson and Walsh 1979). It is doubtful that dihydrotestosterone receptor binding or abnormal 5-α reductase activity is the underlying cause of abnormal sexual differentiation observed in Drash syndrome, since in one case analyzed, neither was impaired (McCoy et al. 1983). Several Denys-Drash patients have been reported with gonadoblastomas (a sex steroid-secreting neoplasm composed of germ cells, sex cord structures, and stromal elements) (Eddy and Mauer 1985; Manivel et al. 1987). This may be secondary, since the incidence of gonadal tumors in male pseudohermaphrodites seems to be no greater than in individuals with undescended testes (Scully 1970).

Karyotype analyses in Denys-Drash syndrome patients, when reported, are usually 46 XY with no structural or numerical abnormalities. There has been one report of a 46 XY/XX mosaic (Denys et al. 1967). To our knowledge, there have been only two WT karyotypes from Denys-Drash syndrome patients reported (McCoy et al. 1983; Jensen et al. 1989). No consistent banding abnormalities were observed in the tumor karyotype; specifically, there were no reports of 11p alterations. Jadresic et al. (1990) have reported a 46 XX female with a constitutional deletion of 11p13 who had nephropathy, bilateral WT, aniridia, and mental retardation. This patient provides a clinical overlap between the Denys-Drash and the WAGR syndromes.

The Perlman syndrome

A possible fourth WT-associated syndrome was originally described by Liban and Kozenitzky (1970) and Perlman et al. (1973). These authors reported five children of Jewish-Yemenite second-cousin parents with a disorder manifested by fetal gigantism, renal hamartomas, and WT. Approximately one third of affected individuals develop WT. There are some similarities between this syndrome and BWS. The molecular basis of Perlman syndrome remains to be elucidated.

TUMOR SUPPRESSOR EXPERIMENTS

Somatic cell fusion experiments have proved to be a useful technique for initial studies of the genetic analysis of human malignancy. When malignant human cells are fused to normal human diploid fibroblasts, the neoplastic phenotype is often suppressed. This suppressive activity has been attributed to the chromosomes derived from the normal parent in the fusion. Similar experiments have been performed with a WT cell line and chromosome 11 to show functionally the presence of a tumor suppressor locus (or loci) at 11p (Weissman et al. 1987). A hypoxanthine phosphoribosyltransferase-deficient (HPRT$^-$) variant of the WT cell line, G401, was used as recipient in these experiments. Microcell hybrids were generated by transferring a single copy of human t(X;11) chromosome (11pter>11q23::Xq26>Xqter) from a mouse cell line carrying only this human chromosome. Since the *HPRT* gene had been translocated to chromosome 11, HAT (hypoxanthine, aminopterin, thymidine) selection could be used to select G401 cells that had retained t(X;11) after transfer. Tumorigenicity, as assayed by inoculation of the cells in nude mice, was found to be suppressed in microcell hybrids that retained t(X;11). Only chromosome 11 was capable of imparting this suppressive activity, since transferring chromosome 13 to G401 did not have this effect. Using their somatic cell hybrids, Weissman et al. (1987) were able to demonstrate that no clear relationship existed between *myc* expression with respect to neoplastic behavior. In their experiments, both the tumorigenic WT cell line, G401, and the nontumorigenic microcell hybrids exhibited the same pattern of expression. These findings argue that N-*myc* overexpression in many WTs (Nisen et al. 1986) may be a tumor epiphenomenon.

Oshimura et al. (1990) have demonstrated that the function of the putative suppressor gene(s) on chromosome 11 is effective only in specific tumor cell lines. Introduction of a normal chromosome 11 into a renal cell carcinoma (YCR-1) or a rat ethylnitrosourea-induced nephroblastoma (ENU-T1) failed to suppress tumorigenicity, whereas fusion into a uterine cervical carcinoma (SiHa) and a rhabdomyosarcoma (A204) did. Thus, cell fusion studies support the view that at least one tumor suppressor gene is located on the short arm of chromosome 11.

CHROMOSOME 11p CHANGES IN WILMS' TUMORS

Loss-of-heterozygosity (LOH) studies of WTs serve to highlight the complexities of the genetic changes that contribute to Wilms' tumorigenesis. Initial studies demonstrated LOH at 11p15 in 12 sporadic WTs analyzed (Fearon et al. 1984; Koufos et al. 1984; Orkin et al. 1984; Reeve et al. 1984). Unfortunately, DNA probes from the rest of chromosome 11 were not tested in these studies. A more informative screen of 14 different WTs using probes for 11p15, 11p13, and 11q detected LOH in five tumors (Mannens et al. 1988). In three of these five tumors, LOH was limited to markers within 11p15.5. These results suggest that not only 11p13, but also 11p15.5 is involved in WT development. LOH studies from Reeve et al. (1989), as well as the finding of a ring chromosome, r(11)(p15.5q25), in an individual who subsequently developed WT (Romain et al. 1983), are consistent with this conclusion. Mutations of loci in band 11p15 and 11p13 (*WT1*) may represent separate pathways in Wilms' tumorigenesis. Alternatively, in some WTs, inactivation of both loci may be required, as suggested by some patients with WAGR syndrome whose tumors show LOH restricted to 11p15 (Henry et al. 1989a).

Recent pathological analyses of WT by Beckwith et al. (1989) are consistent with the involvement of several genes in the etiology of WT. Patients with aniridia or Drash syndrome show localization of nephrogenic rests within kidney lobules (intralobar). Individuals with BWS show nephrogenic rests surrounding kidney lobules (perilobar). According to Beckwith et al. (1989), the intralobar distribution of nephrogenic rests implies an earlier developmental disturbance than perilobar rests. It would be interesting to elucidate the status (i.e., nonmutated or mutated) of the 11p13 *WT1* gene (see below) in these different nephrogenic rests.

The existence of a third WT locus has been postulated. Several familial cases of WT have been described in which linkage between WT and DNA markers on the short arm of chromosome 11 both in 11p15 and 11p13 appears to be excluded (Grundy et al. 1988; Huff et al. 1988; C.E. Schwartz et al., in prep.). Reduction to homozygosity at the Ha-*ras*-1 allele (see Fig. 2 for map position) was demonstrated in a tumor of one such familial case (CALC and CAT [see Fig. 2] were also tested, but were not reduced) (Grundy et al. 1988). It would be interesting to determine the status of the WT gene at 11p13 (see below) in such tumors. None of the patients in these familial cases of WT have the extrarenal malformations associated with WAGR syndrome (Cordero et al. 1980; Grundy et al. 1988; Huff et al. 1988).These results raise the interesting possibility that different loci may be implicated in different WTs. These loci may be involved in different steps of the same regulatory pathway, the abrogation of which could lead to the malignant state.

A recent study by Mannens et al. (1990) tested a total of 44 WTs for LOH at a number of other tumor suppressor regions (for review, see Bishop 1991). In 11 of 36 informative tumors, LOH occurred for DNA markers only on 11p, except for one tumor showing additional LOH for regions 5q, and 17p. No LOH was found for 3p, 13q, and 22q. Thus, LOH studies do not suggest the location of other recessive oncogenes that may contribute to WT etiology. Transfection of WT DNA into NIH-3T3 cells failed to demonstrate the presence of dominant transforming genes in WTs (Orkin et al. 1984).

LOH studies have revealed an interesting pattern with respect to parental chromosome loss during development of WT (Orkin et al. 1984; Reeve et al. 1984; Schroeder et al. 1987; Mannens et al. 1988; Dao et al. 1989; Williams et al. 1989; Huff et al. 1990). In all informative cases, with the exception of one (Huff et al. 1990), the alleles lost in the tumors are of maternal origin. This result has been interpreted to suggest that imprinting of a WT locus on chromosome 11 may contribute to the etiology of the tumor (Scrable et al. 1989). Alternatively, these results may indicate that a preponderance of constitutional mutations to WT suppressor gene(s) occurs during male gamete formation.

DEVELOPING A PHYSICAL MAP OF CHROMOSOME 11

The development of a physical and genetic map of the short arm of chromosome 11 played a crucial role in the identification and localization of some of the genes contributing to the WAGR phenotypes. Somatic cell hybrids have made a particularly crucial contribution in the molecular dissection of 11p.

11p somatic cell hybrids

One of the most important sets of reagents developed for chromosome 11 mapping has been derived from hybrid cell line J1. J1, a human-hamster somatic cell hybrid that stably maintains chromosome 11 as its only human DNA, has served as the parent in generating a series of cell lines that retain overlapping segments of chromosome 11. The J1 hybrid cell line had its origin in 1971 when Puck et al. (1971) generated a series of somatic cell hybrids by fusing human aminocytes with an auxotrophic Chinese hamster cell line (CHO-K1). J1, a subclone of the original fusion, was found to stably maintain chromosome 11 despite years of passage in the apparent absence of selection (Kao et al. 1976). The unusual stability of chromosome 11 in J1 is due to complementation of a mutation present in the CHO-K1 fusion partner by a gene that is tightly linked to the *INS* and *HRAS1* genes on chromosome 11. J1 deletion segregants were isolated by mutagenesis and negative selection for *MIC1*, a cell-surface glycoprotein that can be targeted using monoclonal antibodies (Kao et al. 1976). All J1 deletion derivatives retained at least the distal portion of 11p and are nested contiguous deletions that span the *MIC1* gene whose product was originally selected against. A deletion map of 11p was constructed based solely on the pattern of marker segregation among an expanded panel of J1 deletion derivatives (Glaser 1988). This map (Fig. 2) was consistent both internally and with independently obtained mapping data for 11p. It subdivides 11p into over 20 segments, ordering 35 breakpoints and 36 genetic markers into intervals of average size of less than 2000 kb. The J1 hybrid mapping panel has contributed significantly to the development of a fine structure map of 11p. First, it provides a means to localize rapidly any gene or genetic marker to a defined interval on 11p. Second, it has been used to assess the extent of deletions and/or translocations associated with WAGR or familial aniridia patients' chromosomes 11 (discussed below). Third, individual clones within the mapping panel have been used to generate radiation-reduced hybrids as a means of isolating specific, subchromosomal fragments of 11p (discussed below). Finally, genomic libraries of individual J1 series hybrids have produced many new 11p-specific markers including several that map to the WAGR region (Glaser 1988).

Physical linkage between 11p markers has also been demonstrated by chromosome-mediated gene transfer (CMGT). CMGT is a technique that enables subchromosomal lengths of DNA from one species to be effectively isolated on the background of another. Porteous et al. (1987) have used CMGT specifically to isolate subchromosomal fragments of human 11p within the context of the mouse genome. Mitotic chromosomes from the human EJ bladder carcinoma cell line were used to transform murine C127 cells by calcium-phosphate-mediated coprecipitation. Transformants harboring subchromosomal fragments of 11p were identified by their ability to form foci due to the presence of the actively transforming *HRAS1* oncogene. CMGT has been associated with a relatively high rate of molecular rearrangements and asyntenic cotransfer events. Nevertheless, CMGT has been successfully employed in linkage mapping of several 11p-specific loci through analysis of their pattern of cosegregation over an extensive panel of transformants (Porteous et al. 1987; Bickmore et al. 1988).

A second panel of hybrids has been established that directly complement the J1 series hybrids. These hybrids were constructed by a modification of the irradiation-reduction procedure of Goss and Harris (1975), using cell line J1-11 (a J1 deletion derivative retaining 11p as its only human DNA) as donor and CHO-K1 cells as recipient. Hybrid clones retaining the p13 segment of chromosome 11 were selected for their expression of the *MIC1* surface antigen (Glaser et al. 1990b). Clones that retained large fragments of 11p were identified by their coexpression of both the *MIC1* and *MER2* cell-surface antigens and were eliminated from the panel. A total of seven clones were isolated in this manner. They all carried segments of 11p that spanned *MIC1*, ranging in size from 3200 kb to more than 50,000 kb. Individual clones have been superimposed on the J1 series hybrid map (Fig. 2) and assessed for physical integrity by in situ hybridization, pattern of marker segregation, and fingerprinting by conventional and pulsed-field gel electrophoresis (PFGE) (Glaser et al. 1990b). All clones were found to be generally stable and with an internal molecular rearrangement that was greatly limited in comparison to CMGT. One of these clones, Goss Harris No. 3A (GH3A), stably maintains a single 3200-kb segment of 11p13 that envelops the WAGR region. The reduced amount of 11p DNA carried by each of these hybrids makes them especially useful for probe isolation/regionalization and PFGE mapping.

11p13 mapping

Genomic libraries have been prepared from DNA isolated from several of the J1 deletion and Goss-Harris hybrids (Gusella et al. 1982; Glaser 1988; Call et al. 1990), as well as from other hybrids harboring chromosome 11 (Bruns et al. 1987; Porteous et al. 1987; Compton et al. 1988; Davis et al. 1988). Human clones were identified by hybridization with human repetitive sequence elements (Gusella et al. 1980). To date, more than 325 markers have been mapped to 11p (Junien and McBride 1989), making it one of the most densely marked of the human autosomal regions.

The ordering of 11p-specific markers, especially within and around the WAGR locus, has also been facilitated by the establishment of cell lines and sets of somatic cell hybrids derived from affected patients. The catalase gene (*CAT*) was found to be the first useful marker for the WT region. Junien et al. (1980) have correlated the level of erythrocyte catalase activity with the dosage of band 11p13. Most WAGR patients with a deletion affecting one 11p homolog have reduced levels of this enzyme. Subsequent molecular studies showed that the *CAT* gene is deleted in these patients (Bruns et al. 1984; Michalopoulos et al. 1985; Schroeder et al. 1985; van Heyningen et al. 1985; Glaser et al. 1986). Subsequently, Glaser et al. (1986) found that the gene encoding the β-subunit of follicle-stimulating hormone was deleted in patients with aniridia and WT. These data, coupled with that for the *CAT* gene, bracketed the WAGR complex to an estimated region of 3000–6000 kbp (see Fig. 2).

Several panels of somatic cell hybrids have been generated that isolate the chromosomes 11 from various WAGR and familial aniridia patients. The extent of the overlapping structural abnormalities (i.e., deletion and/or translocation) corresponding to each patient's chromosome 11 has been established by analyzing the segregation pattern of DNA markers within these hybrids (Davis et al. 1988, 1990; Couillin et al. 1989; Gessler et al. 1989). Data on marker order were generated with a somatic cell hybrid panel described by Davis et al. (1988). This panel includes human-mouse somatic cell hybrids that retained an intact human chromosome 11, human-mouse somatic cell hybrids that did not retain human chromosome 11, and several cell lines that harbored deletion or translocation chromosomes 11 from various WAGR and familial aniridia patients. An 11p map generated with this hybrid panel localizes over 50 anonymous genomic clones to nine separable groups. Of these clones, nine were mapped to 11p13, and two of the nine were found to distinguish the WT and aniridia loci. Taken together, these data enabled deduction of the relative order for the genes both within and around the WAGR complex, cen...*CAT-MIC1*-WT-aniridia-*FSHB*...tel, and permitted the positioning of individual anonymous DNA sequences within the region.

Pulsed-field gel electrophoresis mapping

PFGE restriction maps of subchromosomal regions provide the finest resolution physical map of large regions of DNA. Two complementary strategies have been employed in the mapping of 11p13 by PFGE (Compton et al. 1988; Gessler and Bruns 1989; Rose et al. 1990). First, a series of markers previously mapped to 11p13 were localized to individual restriction fragments by hybridization to DNA samples from WAGR region deletion and translocation patients as well as normal controls (Davis et al. 1988; Gessler and Bruns 1989). The relative order of markers could be determined in many cases by cohybridization to specific restriction fragments, analysis of partial or double digests, and determination of restriction fragments interrupted by translocation breakpoints. However, the ability to establish an unambiguous, continuous PFGE map was limited by the number of available markers. A complementary strategy employed by Rose et al. (1990) involved the analysis by PFGE of DNAs prepared from the Goss-Harris hybrid panel (described above) digested to completion with single rare-cutting restriction enzymes, separated by PFGE, and hybridized to human interspersed repetitive DNA. In this way, it was possible to identify the complete complement of restriction fragments located in the WAGR region; 11p13-specific probes, independently ordered by their segregation pattern within the J1 hybrid panel, could then be assigned to individual WAGR region restriction fragments examined by Southern hybridization. By superimposing this map onto the J1-deletion and Goss-Harris hybrid maps, a complete physical map of 11p13 was obtained (Rose et al. 1990). All three physical maps have been compared with hybrid mapping panels that segregate the chromosomes 11 of various WAGR and familial aniridia patients. Such analysis has enabled localization of the aniridia and WT loci within the WAGR region to *Not*I fragments of 1380 kb (Compton et al. 1988, Gessler and Bruns 1989; Rose et al. 1990) and 345 kb (Rose et al. 1990), respectively.

WT1 ISOLATION AND FUNCTIONAL STUDIES

The elucidation of a physical map of 11p13 provided a means to focus the search for potential candidates of the 11p13 WT gene. Of particular importance in screening these candidate clones was a sporadic WT, Wit 13, identified with homozygous deletion of genetic material in band 11p13 (Lewis et al. 1988). The region of homozygous deletion in this sporadic tumor proved to be within the boundaries for the WT gene locus previously identified from WAGR patients. Assuming that this tumor had acquired its transformed phenotype by deleting both copies of the 11p13 WT gene, then candidate genes must fall within the smallest region of overlap (SRO) of the deletions, estimated to be approximately 350 kbp (see Fig. 3) (Rose et al. 1990). One cosmid clone, J8-3, that fell within this SRO (see Fig. 3) was further characterized because of the high homology with genomic DNA of other mammalian species exhibited by one subclone, J8-3p4. J8-3 represents a segment of a gene that spans approximately 50 kbp and whose transcription proceeds in a centromeric to telomeric direction (see Fig. 3) (Call et al. 1990). Gessler et al. (1990), using a chromosome jumping approach, isolated a *Not*I linking clone that represents the HTF island at the 5' end of this gene.

Several cDNAs representing the transcription (which we refer to as *WT1*) were isolated and sequenced. The predicted polypeptide encoded by the *WT1* gene shows a number of characteristics of a transcription factor. It contains four zinc fingers of the Cys-His variety at the carboxyl

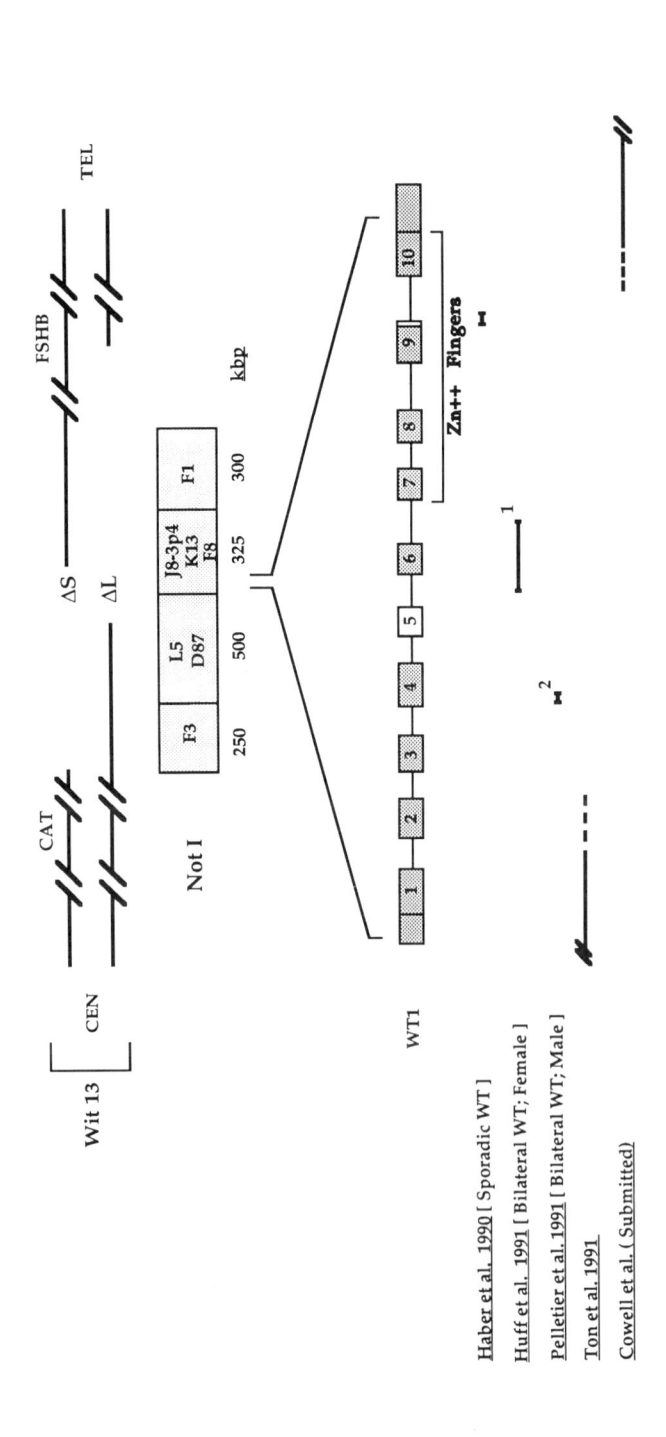

Haber et al. 1990 [Sporadic WT]
Huff et al. 1991 [Bilateral WT; Female]
Pelletier et al. 1991 [Bilateral WT; Male]
Ton et al. 1991
Cowell et al. (Submitted)

[1] (Both Tumors Homozygous; Patient is Heterozygous)

[2] (Both Tumors Homozygous; Patient is Heterozygous).

Figure 3 Schematic representation of the chromosomal location and structure of *WT1*. The pulsed-field *Not*I map of the WAGR region was determined by Rose et al. (1990) and has been superimposed on the homozygous deletion detected in the sporadic WT cell line, Wit13. The map positions of several genomic clones isolated by Glaser (1988), as well as J8-3p4 (described in Call et al. 1990) are shown. *WT1* contains 10 exons, two of which are alternative splice choices (denoted by white boxes) (D.A. Haber et al., in prep.). The exons are not drawn to scale. Deletions that have been detected in sporadic and hereditary cases of WT are shown below the exon/intron structure.

terminus, as well as a proline/glutamine-rich amino terminus (Call et al. 1990; Gessler et al. 1990). Such proline/glutamine-rich regions are found in other DNA-binding proteins, including Krüppel and the CTF/NF-1 family of closely related polypeptides (Mitchell and Tjian 1989). Three of the four zinc finger domains of *WT1* show a 51–63% homology with the three zinc fingers of the early growth response (EGR) genes 1 and 2 (Call et al. 1990; Gessler et al. 1990). These proteins are induced when serum-starved fibroblasts are stimulated by the addition of fresh serum to enter G_1 (Joseph et al. 1988). The *WT1* gene product binds a DNA sequence similar to the EGR recognition site, as recently demonstrated by Rauscher et al. (1990). The EGR proteins are thought to be important nuclear intermediates in the signal transduction pathway for growth response. The homology and conservation in DNA binding between *WT1* and the EGRs suggested to us that *WT1* may be involved in the growth proliferation response of nephroblasts and that this regulation had been uncoupled in nephroblastomas. Using immunofluorescence, we have shown that *WT1* localizes to the nucleus after introduction into COS-1 cells, consistent with its predicted role as a transcription factor (Pelletier et al. 1991a).

The zinc finger amino acid sequence of *WT1* is also closely related to the yeast gene product, *MIG1*. The similarity is particularly pronounced in the fingertip of the loops, which is thought to be the region important for DNA binding (Nardelli et al. 1991). *MIG1* has been demonstrated to regulate transcription levels of genes involved in sugar metabolism (Nehlin and Ronne 1990). The *MIG1* recognition site is similar to the EGR- and *WT1*-binding sites. The similarities between *WT1* and *MIG1* suggest that *WT1*, like *MIG1*, may function as a repressor. This would be consistent with its proposed role in development and tumorigenesis.

Northern blot analysis demonstrated that the *WT1* cDNA recognized an approximately 3-kb mRNA whose expression was highest in kidney and spleen (Call et al. 1990; Gessler et al. 1990). Four alternatively spliced mRNA species are synthesized due to the presence of two alternative splice choices in the *WT1* pre-mRNA (see Fig. 3) (Haber et al. 1990 and in prep.). There is no detectable expression in brain, intestine, liver, muscle, lung, thymus, skin, stomach, or eye (Call et al. 1990; Gessler et al. 1990; Pritchard-Jones et al. 1990). *WT1* RNA expression is very abundant in human fetal kidney but not detectable in adult kidney (Haber et al. 1990). In addition, *WT1* mRNA levels were examined in several tumor cell lines and found to be present in K562 (erythroleukemia), CEM (acute lymphocytic leukemia) (Call et al. 1990), and 293 cells (adenovirus-transformed human embryonic kidney cell line) (Pritchard-Jones et al. 1990).

Pritchard-Jones et al. (1990) have elegantly defined *WT1* mRNA expression during kidney development by in situ RNA hybridizations on fetal kidney sections. The *WT1* gene is expressed during nephrogenesis in the condensing blast cells and the renal vesicle and then is restricted to the developing podocyte cells of the glomerular epithelium (Pritchard-Jones et al. 1990). No expression is seen in the proximal and distal tubules or in derivatives of the ureteric bud. Expression was also observed in the glomerular epithelium of the mesonephros, which functions as an excretory system for a short period of time during embryogenesis in humans (Pritchard-Jones et al. 1990).

Highly variable levels of *WT1* RNA are observed in sporadic WTs (Haber et al. 1990; Pritchard-Jones et al. 1990). mRNA transcripts with altered mobility, indicative of a large intragenic deletion or translocation, are rarely observed by Northern blotting. Some WTs contain deletions that encompass the *WT1* gene, along with adjacent genomic DNA sequences (Call et al. 1990; Gessler et al. 1990). However, several tumors were identified in which the region of homozygous deletion specifically involved the *WT1* transcription unit. These tumors show deletions extending into the upstream exons (Ton et al. 1991) or into the downstream exons of *WT1* (see Fig. 3) (J. Cowell et al., in prep.). Small internal deletions that directly alter mRNA structure have also been identified within *WT1* in WT (Haber et al. 1990; Huff et al. 1991; Pelletier et al. 1991b). These findings support the identification of this transcription unit and its gene product as the 11p13 WT gene. The status of *WT1* in nephrogenic rests, which are thought to be precursor lesions of WT, has not yet been established.

At least one mutation in a WT has the characteristics of a so-called dominant-negative mutation. A 25-bp deletion at the 3′ splice junction of exon 9 causes the removal of the nucleotide sequences encoding exon 9 from the *WT1* mRNA to yield an in-frame deletion (Haber et al. 1990). The polypeptide encoded by this mRNA is predicted to be lacking zinc finger 3. This polypeptide fails to bind DNA (Rauscher et al. 1990) but may still interact competitively with protein targets of the *WT1* polypeptide. Such a situation would allow a mutated gene product to competitively inhibit function of the wild-type *WT1* product, thus creating a dominant-negative effect. This issue awaits further analysis by site-directed mutagenesis of the *WT1* gene, as well as the development of functional assays for the *WT1* gene product.

The *WT1* gene is expressed at high levels in the differentiating testes and ovaries (Pelletier et al. 1991a). In contrast to the kidney, expression in these organs continues at high levels throughout adult life. In addition to the role of *WT1* in kidney development, these observations implicate *WT1* in additional developmental pathways, namely, ovarian and testicular development. The most common genital anomalies among individuals with WT-associated syndromes are hypospadias and cryptorchidism (Miller et al. 1964; Drash et al. 1970). GU defects are more often associated with bilateral WT than is aniridia (Pendergrass 1976). Several recent studies have suggested that the *WT1* gene product itself is involved in GU development. (1) *WT1* is expressed in the GU system. The main sites of expression are the fetal gonad, genital ridge, ovary, testis (Pritchard-Jones et al. 1990; Pelletier et al. 1991a), and uterus (Pelletier et al. 1991a). (2) Genetic analysis using restriction-fragment-length polymorphism (RPLP) and dosage analysis of 36 genetic markers failed to reveal any rearrangements or deletions of 11p13 in a child with bilateral WT and hypospadias (patient PG) (Glaser et al. 1989). This child carries a 17-bp deletion that disrupts an exon upstream of the zinc finger domains and creates an in-frame stop codon. This mutation is converted to homozygosity in both his WTs, which presumably initiated malignant transformation (see Fig. 3) (Pelletier et al. 1991b). (3) An association between urogenital anomalies and early-onset WTs has also been suggested on the basis of statistical data (Breslow et al. 1988; Beckwith et al. 1989). GU anomalies would represent the hemizygous expression of mutations within *WT1*. This suggests that GU anomalies in WAGR individuals are a dominant phenotype of null mutations to *WT1* and underscore the sensitivity of the developing urogenital system to *WT1* gene dosage. Further reduction of *WT1* expression could occur in hemizygous individuals if the normal allele is subject to germ line imprinting or positive feedback regulation. This raises the interesting possibility that somatic hemizygous inactivation of *WT1* may be quite frequent and may result in apparently non-WT-associated GU anomalies.

We have recently analyzed the status of the *WT1* gene in ten individuals with Denys-Drash syndrome. We find that these individuals contain specific germ-line point mutations within the zinc finger domains of one *WT1* allele (Pelletier et al. 1991c). These mutations directly affect DNA sequence recognition and are thought to act in a dominant-negative manner. These mutations provide genetic evidence for a key role of *WT1* in urogenital development and provide the first example of a recessive oncogene implicated in a human malformation syndrome.

A separate genitourinary dysplasia locus (GUD) in band 11p13 has been suggested (Porteous et al. 1987; Bickmore et al. 1989) on the basis of a male neonate with clinical features of the Potter (or oligohydramnios) sequence (Thomas and Smith 1974), several urogenital malformations (urethral and bilateral ureteral atresia, cryptorchidism), and a balanced t(2;11)(p11;p13) translocation. Recent molecular analysis of the WAGR region indicates that this translocation breakpoint on 11p is separated from the *WT1* locus by more than 650 kbp (Rose et al. 1990; van Heyningen et al. 1990). It is possible that this translocation is unrelated to the patient's urogenital defects or that it decreases *WT1* gene expression through a position (*cis*) effect (Lewis 1950), since the patient's phenotypically normal mother carries the identical translocation (H. Punnett, pers. comm.), and the breakpoint on chromosome 2 is adjacent to a region of centromeric heterochromatin (band 2p11). The documented expression of *WT1* in the urogenital system (Pritchard-Jones et al. 1990; Pelletier et al. 1991a) suggests that the *WT1* gene product is directly involved in genitourinary development.

WT1 expression is observed at lower levels in a number of other adult organs, including spleen, heart, and lung (Call et al. 1990; Buckler et al. 1991). The role that the *WT1* gene product plays in these tissues is unclear. It is of interest to note that *WT1* is expressed in some leukemic cell lines (D.A. Haber, unpubl.). Whether lesions in the *WT1* gene play a role in malignancies other than WT remains an intriguing question.

MOUSE *Sey* LOCUS

Kidney tumors occur relatively frequently in rats. In 1950, Olcott succeeded in transplanting a nephroblastoma observed in a male rat through three passages in Wistar rats. Subsequently, Thompson et al. (1961) described a spontaneous nephroblastoma in a Sprague-Dawley 8-month-old female. These tumors can also be induced by chemical agents or radiation in the rat (for review, see Guerin et al. 1969). The Wistar/Furth rat develops a spontaneous WT that has been used as a model for chemotherapy and radiotherapy (Tomashefsky et al. 1976).

A strikingly similar disease can also be produced in chickens with an avian myeloblastosis virus derivative (Watts and Smith 1980) or in opossums fed with ethylnitrosourea (Jurgelski et al. 1976). The nephroblastomas observed in these systems arise from the same cell type as WT. However, they do not accurately reflect the human disease, which can be hereditary and results from a loss of a gene.

Primitive kidney tumors are extremely rare in mice. The most important work on this subject has been by Slye et al. (1921) who performed 33,000 autopsies on mice and reported only five renal epithelial tumors. The first spontaneous WT noted in a mouse was reported by Guerin et al. (1969).

A mouse deletion similar to WAGR deletions recently has been described by Glaser et al. (1990a). An interspecies backcross between *Mus musculus/domesticus* and *Mus spretus* was used to map the region in the mouse genome syntenic to human 11p13. Nine evolutionarily conserved DNA markers from proximal 11p were used to score a panel of 94 interspecies backcross mice (Glaser et al. 1990a). These probes cosegregated in the center of mouse chromosome 2, between *ld*, a gene involved in limb development, and the glucagon gene (*Gcg*). The results mapped the murine homolog to the WAGR region close to the Small-eyes (*Sey*) mutation, the phenotype and inheritance pattern of which resemble those of human aniridia (Theiler et al. 1978; Hogan et al. 1986).

The *Sey* gene has been mapped in multipoint crosses approximately 5 cM proximal to *pallid* (*pa*) (Hogan et al. 1986). The Sey^{MH} allele was recently localized between *Fshb* and *Cas-1*. This raises the possibility that *Sey* represents mutations within the murine equivalent of the aniridia gene (van der Meer-de Jong et al. 1990). The most deleterious allele is *Sey^{Dey}*, which arose spontaneously in the C3H/HeJ strain. Heterozygous animals can be identified by reduced size of the eyes, a 10% reduction in body size, and on most, but not all affected animals, a small white belly spot (Theiler et al. 1978). Fetal loss among heterozygotes is about 60%, and homozygotes die on or before the sixth day of gestation (Theiler et al. 1978). Interestingly, carriers of the human aniridia trait transmit the disease with a 38:62 ratio rather than the expected 50:50 for a Mendelian dominant gene (Shaw et al. 1960).

Southern blot analysis of *Sey^{Dey}*/+ mice revealed that between approximately 1300 kbp and 2300 kbp is deleted from the *Sey^{Dey}* allele (Glaser et al. 1990a). This deletion encompasses the murine *WT1* gene (Glaser et al. 1990a; Buckler et al. 1991); however, unlike their human counterparts, *Sey^{Dey}*/+ mice do not develop nephroblastomas (Glaser et al. 1990a). This may be due to a smaller and/or more short-lived target cell population in mice than in humans. Although the exact target cell population is hard to determine, mice have approximately 100 times fewer nephrons than humans per adult kidney, consistent with a smaller original metanephric blastema cell population (Potter 1972). This would effectively decrease the window of opportunity for somatic mutations to inactivate the remaining normal allele. In addition, WTs are thought to arise in utero (Potter 1972). Therefore, the shorter gestational period in mice also decreases the probability for somatic mutations. These differences are predicted to reduce significantly the incidence of nephroblastomas among *Sey^{Dey}*/+ mice. Unlike their human counterparts, *Sey^{Dey}*/+ mice do not have genitourinary abnormalites. This suggests that the underlying developmental events in mice are less sensitive to reduced levels of *WT1*.

The isolation and characterization of a series of cDNA clones from the mouse corresponding to the human *WT1* cDNA have been reported (Buckler et al. 1991). An extremely high degree of identity is observed in the deduced amino acid sequence between the gene products of the two species. The identity extends throughout the entire coding region of the molecule. The mouse model system and *WT1* cDNAs should prove useful reagents for elucidating some of the biochemical events involved in gonadal and kidney organogenesis.

CONCLUSION

The challenges presented by WT have offered important opportunities in the study of tumorigenesis. The identification of *WT1* is the first step in a long process aimed at dissecting the steps that result in WT formation.

Characterization of *WT1*-encoded protein function should provide a link between the regulation of transcription, cellular proliferation, and cellular differentiation. The involvement of additional loci in WT etiology provides a framework for an incisive genetic and biochemical analysis of this gene(s). It is hoped that these studies will provide us with more detailed information regarding the control of cellular proliferation in more complex cancers.

References

Aron, B. 1974. Wilms' tumor—A clinical study of eighty-one children. *Cancer* **33**: 637.

Barakat, A.Y., Z.L. Papadopoulou, R.S. Chandra, C.E. Hollerman, and P.L. Calcagno. 1974. Pseudohermaphroditism, nephron disorder and Wilms' tumor: A unifying concept. *Pediatrics* **54**: 366.

Beckwith, J.B. 1963. Extreme cytomegaly of the fetal adrenal cortex, omphalocoele, hyperplasia of kidneys and pancreas, and Leydig-cell hyperplasia. Another syndrome? Presented at the Annual Meeting of Western Society for Pediatric Research, Los Angeles, California, Nov. 11.

Beckwith, J.B., N.B. Kiviat, and J.F. Bonadio. 1989. Nephrogenic rests, nephroblastomatosis, and the pathogenesis of Wilms' tumor. *Pediatr. Pathol.* **10**: 1.

Bell, G.I., D.S. Gerhard, N.M. Fong, R. Sanchez-Pescador, and L.B. Rall. 1985. Isolation of the human insulin-like growth factor genes: Insulin-like growth factor II and insulin are contiguous. *Proc. Natl. Acad. Sci.* **82**: 6450.

Bennington, J.L. and J.B. Beckwith. 1975. Tumors of the kidney, renal pelvis and ureter. In *Atlas of tumor pathology*, series 2, fascile 12. Armed Forces Institute of Pathology, Washington, D.C.

Bickmore, W., S. Christie, V. van Heyningen, N.D. Hastie, and D.J. Porteous. 1988. Hitch-hiking from HRAS1 to the WAGR locus with CMGT markers. *Nucleic Acids Res.* **16**: 51.

Bickmore, W.A., D.J. Porteous, S. Christie, A. Seawright, J.M. Fletcher, J.C. Maule, P. Couillin, C. Junien, N.D. Hastie, and V. van Heyningen. 1989. CpG islands surround a DNA segment located between translocation breakpoints associated with genitourinary dysplasia and aniridia. *Genomics* **5**: 685.

Bishop, J.M. 1991. Molecular themes in oncogenesis. *Cell* **64**: 235.

Bond, J.V. 1975. Bilateral Wilms' tumor. Age at diagnosis, associated congenital anomalies, and possible pattern of inheritance. *Lancet* **II**: 482.

Bove, K.E., H. Koffler, and A.J. McAdams. 1969. Nodular renal blastema. Definition and possible significance. *Cancer* **24**: 323.

Breslow, N.E. and J.B. Beckwith. 1982). Epidemiological features of Wilms' tumor: Results of the National Wilms' Tumor Study. *J. Natl. Cancer Inst.* **68**: 429.

Breslow, N.E., J.B. Beckwith, M. Ciol, and K. Sharples. 1988. Age distribution of Wilms' tumor: Report from the National Wilms' Tumor Study. *Cancer Res.* **48**: 1653.

Broduer, G.M. 1984. Genetic and cytogenetic aspects of Wilms' Tumor. In *Wilms' tumor: Clinical and biological manifestations* (ed. C. Pochedly and E.S. Baum), p. 125. Elsevier Science, New York.

Brown, K.W., Williams, J.C., N.J. Maitland, and M.G. Mott. 1990. Genomic imprinting and the Beckwith-Wiedemann syndrome. *Am. J. Hum. Genet.* **46:** 1000.

Bruns, G.A.P., S.D. Barnes, M. Gessler, J.B. Brennick, and M.J. Weiner. 1987. DNA probes for chromosome 11 and the WAGR deletion. *Cytogenet. Cell Genet.* **46:** 588.

Bruns, G.A.P., T. Glaser, J.F. Guesella, D.E. Housman, and S. Orkin. 1984. Chromosome 11 probes identified with a catalase oligonucleotide. *Am. J. Hum. Genet.* **36:** 255.

Buckler, A.J., J. Pelletier, D.A. Haber, T. Glaser, and D.E. Housman. 1991. Isolation, characterization, and expression of the murine Wilms' tumor gene (WT1) during kidney development. *Mol. Cell. Biol.* **11:** 1707.

Call, K.M., T. Glaser, C.Y. Ito, A.J. Buckler, J. Pelletier, D.A. Haber, E.A. Rose, A. Kral, H. Yeger, W.H. Lewis, C. Jones, and D.E. Housman. 1990. Isolation and characterization of a zinc finger polypeptide gene at the human chromosome 11 Wilms' tumor locus. *Cell* **60:** 509.

Cordero, J.F., F.P. Li, L.B. Holmes, and P.S. Gerald. 1980. Wilms' tumor in five cousins. *Pediatrics* **66:** 716.

Compton, D.A., M.M. Weil, C.A. Jones, V.M. Riccardi, L.C. Strong, and G.F. Saunders. 1988. Long range physical map of the Wilms' tumor-aniridia region on human chromosome 11. *Cell* **55:** 827.

Cotlier, E., M. Rose, and S.A. Moel. 1978. Aniridia, cataracts, and Wilms' tumor in monozygous twins. *Am. J. Ophthalmol.* **86:** 129.

Couillin, P., M. Azoulay, I. Henry, N. Ravisé, M.C. Grisard, C. Jeanpierre, F. Barichard, F. Metezeau, J.J. Chandelier, W. Lewis, V. van Heyningen, and C. Junien. 1989. Characterization of a panel of somatic cell hybrids for subregional mapping along 11p and within band 11p13. Subdivision of the WAGR complex region. *Hum. Genet.* **82:** 171.

Cox, D. 1966. Chromosome constitution of nephroblastomas. *Cancer* **19:** 1217.

D'Angio, G.J., N. Breslow, J.B. Beckwith, A. Evans, E. Baum, A. DeLorimier, D. Fernbach, E. Hrabovsky, B. Jones, P. Kelalis, H.B. Othersen, M. Tefft, and P.R.M. Thomas. 1989. Treatment of Wilms' tumor: Results of the Third National Wilms' Tumor Study. *Cancer* **64:** 349.

Dao, D.D., W.T. Schroeder, L.-Y. Chao, H. Kikuchi, L.C. Strong, V.M. Riccardi, S. Pathak, W.W. Nichols, W.H. Lewis, and G.F. Saunders. 1987. Genetic mechanisms of tumor-specific loss of 11p DNA sequences in Wilms' tumor. *Am. J. Hum. Genet.* **41:** 202.

Davis, L.M., M.G. Byers, Y. Fukushima, S. Qin, N.J. Nowak, C. Scoggin, and T.B. Shows. 1988. Four new DNA markers are assigned to the WAGR region of 11p13: Isolation and regional assignment of 112 chromosome 11 anonymous DNA segments. *Genomics* **3:** 264.

Davis, L.M., G. Senger, H.-J. Lüdecke, U. Claussen, B. Horsthemke, S.S. Zhang, B. Metzroth, K. Hohenfellner, B. Zabel, and T.B. Shows. 1990. Somatic cell hybrid and long-range physical mapping of 11p13 microdissected genomic clones. *Proc. Natl. Acad. Sci.* **87:** 7005.

DeChiara, T.M., E.J. Robertson, and A. Efstratiadis. 1991. Parental imprinting of the mouse insulin-like growth factor II gene. *Cell* **64:** 849.

Denys, P., P. Malvaux, H. van den Berghe, W. Tanghe, and W. Proesmans. 1967. Association d'un syndrome anatomo-pathologique de Pseudohermaphrodisme masculin, d'une tumeur de Wilms', d'une nephropathie parenchymateuse et d'un mosaicisme XX/XY. *Arch. Fr. Pediatr.* **24**: 729.

Diaz de Bustamante, A., A. Delicado, P. Garcia de Miguel, M.T. Darnaude, M.L. de Torres, R.M. Zumel, and I. Lopez Pajares. 1990. Balanced reciprocal translocation (X;20) limited to Wilms' tumor in a Wiedemann-Beckwith syndrome. *Cancer Genet. Cytogenet.* **45**: 35.

Douglass, E.C., J.A. Wilimas, A.A. Green, and A.T. Look. 1985. Abnormalities of chromosomes 1 and 11 in Wilms' tumor. *Cancer Genet. Cytogenet.* **14**: 331.

Drash, A., F. Sherman, W.H. Hartmann, and R.M. Blizzard. 1970. A syndrome of pseudohermaphroditism, Wilms' tumor, hypertension, and degenerative renal disease. *J. Pediatr.* **76**: 585.

Eddy, A.A. and S.M. Mauer. 1985. Pseudohermaphroditism, glomerulopathy, and Wilms' tumor (Drash syndrome): Frequency in end stage renal failure. *J. Pediatr.* **106**: 584.

Fearon, E.R., B. Vogelstein, and A.P. Feinberg. 1984. Somatic deletion and duplication of genes on chromosome 11 in Wilms' tumours. *Nature* **309**: 176.

Ferrell, R.E. and V.M. Riccardi. 1981. Catalase levels in patients with aniridia and/or Wilms' tumor: Utility and limitations. *Cytogenet. Cell Genet.* **31**: 120.

Ferrell, R.E., L.C. Strong, S. Pathak, and V.M. Riccardi. 1983. Wilms' tumor (WT) cytogenetics: The variable presence of del(11p) in WT explants. *Am. J. Hum. Genet.* **35**: 63A.

Francois, J., D. Coucke, and R. Coppieters. 1977. Aniridia-Wilms' tumor syndrome. *Ophthalmology* **174**: 35.

Francke, U., L.B. Holmes, L. Atkins, and V.M. Riccardi. 1979. Aniridia-Wilms' tumor association: Evidence for specific deletion of 11p13. *Cytogenet. Cell Genet.* **24**: 185.

Fraumeni, J.F. and A.G. Glass. 1968. Wilms' tumor and congenital aniridia. *J. Am. Med. Assoc.* **206**: 825.

Gallo, G. and H.E. Chemes. 1987. The association of Wilms' tumor, male pseudohermaphroditism and diffuse glomerular disease (Drash syndrome). Report of eight cases with clinical and morphologic findings and review of the literature. *Pediatr. Pathol.* **7**: 175.

Gardner, R.J.M., R.M. Grindley, W.E. Chewings, and M.D.H. Holdaway. 1983. Wilms' tumor and somatic rearrangements of chromosome 11 at band p13. *Proc. Jpn. Cancer Assoc.* 181. (Abstr.)

Garfunkel, J.M. 1985. Editors comments on Edidin, D.V. Pseudohermaphroditism, glomerulopathy, and Wilms' tumor (Drash syndrome). *J. Pediatr.* **107**: 988.

Geiser, C.F. and A.M. Shindler. 1969. Long term survival in a male with 18-trisomy syndrome and Wilms' tumor. *Pediatrics* **44**: 11.

Gessler, M. and G.A.P. Bruns. 1989. A physical map around the WAGR complex on the short arm of chromosome 11. *Genomics* **5**: 53.

Gessler, M., A. Poustka, W. Cavenee, R.L. Neve, S.H. Orkin, and G.A.P. Bruns. 1990. Homozygous deletion in Wilms' tumours of a zinc-finger gene identified by chromosome jumping. *Nature* **343**: 774.

Gessler, M., G.H. Thomas, P. Couillin, C. Junien, B.C. McGillivray, M. Hayden, G. Jaschek, and G.A.P. Bruns. 1989. A deletion map of the WAGR region

on chromosome 11. *Am. J. Hum. Genet.* **44:** 486.
Gilgenkrantz, S., C. Vigneron, M.J. Gregoire, C. Pernot, and A. Raspiller. 1982. Association of del(11)(p15.1p12), aniridia, catalase deficiency, and cardiomyopathy. *Am. J. Hum. Genet.* **13:** 39
Glaser, T. 1988. *The line structure and evolution of the eleventh human chromosome.* Ph.D. thesis, Massachusetts Institute of Technology, Cambridge.
Glaser, T., J. Lane, and D. Housman. 1990a. A mouse model of the aniridia-Wilms' tumor deletion syndrome. *Science* **250:** 823.
Glaser, T., C. Jones, E.C. Douglass, and D. Housman. 1989. Constitutional and somatic mutations of chromosome 11p in Wilms' tumor. *Cancer Cells* **7:** 253.
Glaser, T., E. Rose, H. Morse, D. Housman, and C. Jones. 1990b. A panel of irradiation-reduced hybrids selectively retaining human chromosome 11p13: Their structure and use to purify the WAGR gene complex. *Genomics* **6:** 48.
Glaser, T., W.H. Lewis, G.A.P. Bruns, P.C. Watkins, C.E. Rogler, T.B. Shows, V.E. Powers, H.F. Willard, J.M. Goguen, K.O.J. Simola, and D.E. Housman. 1986. The β-subunit of follicle-stimulating hormone is deleted in patients with aniridia and Wilms' tumour, allowing a further definition of the WAGR locus. *Nature* **321:** 882.
Goldman, S.M., D.J. Garfinkel, K.S. Oh, and J.P. Dorst. 1981. The Drash syndrome: Male pseudohermaphroditism, nephritis, and Wilms' tumor. *Radiology* **141:** 87.
Goss, S.J. and H. Harris. 1975. New method for mapping genes in human chromosomes. *Nature* **255:** 680.
Gross, R.E and E.B.D. Neuhauser. 1950. Treatment of mixed tumors of kidney in childhood. *Pediatrics* **6:** 843.
Grundy, P., A. Koufos, K. Morgan, F.P. Li, A.T. Meadows, and W.K. Cavenee. 1988. Familial predisposition to Wilms' tumour does not map to the short arm of chromosome 11. *Nature* **336:** 374.
Guerin, M., I. Chouroulinkov, and M.R. Riviere. 1969. Experimental kidney tumors. In *The kidney* (ed. C. Rouiller and A.F. Muller), vol. 2, p. 199. Academic Press, New York.
Gusella, J.F., C. Jones, F.T. Kao, D. Housman, and T.T. Puck. 1982. Genetic fine-structure mapping in human chromosome 11 by use of repetitive DNA sequences. *Proc. Natl. Acad. Sci.* **79:** 7804.
Gusella, J.F., C. Keys, A. Varsanyi-Brenner, F.-T. Kao, C. Jones, T.T. Puck, and D. Housman. 1980. Isolation and localization of DNA segments from specific human chromosomes. *Proc. Natl. Acad. Sci.* **77:** 2829.
Haber, D.A., A.J. Buckler, T. Glaser, K.M. Call, J. Pelletier, R.L. Sohn, E.C. Douglass, and D.E. Housman. 1990. An internal deletion within an 11p13 zinc finger gene contributes to the development of Wilms' tumor. *Cell* **61:** 1257.
Habib, R., C. Loriat, M.C. Gubler, P. Niaudet, A. Bensman, M. Levy, and M. Broyer. 1985. The nephropathy associated with male pseudohermaphroditism and Wilms' tumor (Drash syndrome): A distinct glomerular lesion—Report of 10 cases. *Clin. Nephrol.* **24:** 269.
Hall, J.G. 1990. Genomic imprinting: Review and relevance to human diseases. *Am. J. Hum. Genet.* **46:** 857.

Haselbacher, G.K., J.C. Irminger, J. Zapf, W.H. Ziegler, and R.E. Humbel. 1987. Insulin-like growth factor II in human adrenal pheochromocytomas and Wilms' tumors: Expression at the mRNA and protein level. *Proc. Natl. Acad. Sci.* **84:** 1104.

Henry, I., M. Jeanpierre, F. Barichard, J.L. Serre, J. Mallet, C. Turleau, J. de Grouchy, and C. Junien. 1988. Duplication of HRAS1, INS, and IGF2 is not a common event in Beckwith-Wiedemann syndrome. *Ann. Genet.* **31:** 216.

Henry, I., S. Grandjouan, P. Couillin, F. Barichard, C. Huerre-Jeanpierre, T. Glaser, T. Philip, G. Lenoir, J.L. Chaussain, and C. Junien. 1989a. Tumor-specific loss of 11p15.5 alleles in del11p13 Wilms' tumor and in familial adrenocortical carcinoma. *Proc. Natl. Acad. Sci.* **86:** 3247.

Henry, I., M. Jeanpierre, P. Couillin, F. Barichard, J.-L. Serre, H. Journel, A. Lamouroux, C. Turleau, J. de Grouchy, and C. Junien. 1989b. Molecular definition of the 11p15.5 region involved in Beckwith-Wiedemann syndrome and probably in predisposition to adrenocortical carcinoma. *Hum. Genet.* **81:** 273.

Higurashi, M., K. Ijima, Y. Sugimoto, N. Ishikawa, H. Hoshina, N. Watanabe, and K. Yoneyarna. 1980. The birth prevalence of malformatin syndromes in Tokyo infants. *Am. J. Med. Genet.* **6:** 189.

Hogan, B.L.M., G. Horsburgh, J. Cohen, C.M. Hetherington, G. Fisher, and M.F. Lyon. 1986. *Small eyes (Sey)*: A homozygous lethal mutation on chromosome 2 which affects the differentiation of both lens and nasal placodes in the mouse. *J. Embrol. Exp. Morphol.* **97:** 95.

Horton, C. and C. Devine. 1972. Hypospadias and epispadias. *Clin. Symp.* **24.** (Abstr.)

Huff, V., A. Meadows, V.M. Riccardi, L.C. Strong, and G.F. Saunders. 1990. Parental origin of de novo constitutional deletions of chromosomal band 11p13. *Am. J. Hum. Genet.* **47:** 155.

Huff, V., D.A. Compton, L.-Y. Chao, L.C. Strong, C.F. Geiser, and G.F. Saunders. 1988. Lack of linkage of familial Wilms' tumour to chromosomal band 11p13. *Nature* **336:** 377.

Huff, V., H. Miwa, D.A. Haber, K.M. Call, D.E. Housman, L.C. Strong, and G.F. Saunders. 1991. Evidence for WT1 as a Wilms' tumor (WT) gene: Intragenic germinal deletion in bilateral WT. *Am. J. Hum. Genet.* **48:** 997.

Innis, M.D. 1972. Nephroblastoma: Possible index cancer of childhood. *Med. J. Aust.* **1:** 18.

Irminger, J.C., E.J. Schoenle, J. Briner, and R.E. Humbel. 1989. Structural alteration of the insulin-like growth factor II-gene in Wilms' tumor. *Eur. J. Pediatr.* **148:** 620.

Jadresic, L., J. Leake, I. Gordon, M.J. Dillon, D.B. Grant, J. Pritchard, R.A. Risdon, and T.M. Barrat. 1990. Clinicopathologic review of twelve children with nephropathy, Wilms' tumor, and genital abnormalities (Drash syndrome). *J. Pediatr.* **117:** 717.

Jensen, J.C., R.M. Ehrlich, M.K. Hanna, R.N. Fine, and I. Grunberger. 1989. A report of 4 patients with the Drash syndrome and a review of the literature. *J. Urol.* **141:** 1174.

Joseph, L.J., M.M. Le Beau, G.A. Jamieson, S. Acharya, T. Shows, J.D. Rowley, and V.P. Sukhatme. 1988. Molecular cloning, sequencing, and mapping of EGR2, a human growth response gene encoding a protein with "zinc-binding finger" structure. *Proc. Natl. Acad. Sci.* **85:** 7164.

Journel, H., J. Lucas, C. Allaire, F. Le Mee, G. Defawe, M. Lecornu, H. Jouan, M. Roussey, and B. Le Marec. 1985. Trisomy 11p15 and Beckwith-Wiedemann syndrome. Report of two new cases. *Ann. Genet.* **28:** 97.

Joy Ping, A., A.E. Reeve, D.J. Law, M.R. Young, M. Boehnke, and A.P. Feinberg. 1989. Genetic linkage of Beckwith-Wiedemann syndrome to 11p15. *Am. J. Hum. Genet.* **44:** 720.

Junien, C. and O.W. McBride. 1989. Report on the committee on the genetic constitution of chromosome 11. *Cytogenet. Cell Genet.* **51:** 226.

Junien, C., C. Turleau, J. de Grouchy, R. Said, M.O. Rethore, R. Tenconi, and J.L. Dufier. 1980. Regional assignment of catalase (CAT) gene to band 11p13. Association with the aniridia-Wilms' tumor-gonadoblastoma (WAGR) complex. *Ann. Genet.* **23:** 165.

Jurgelski, W., Jr., P.M. Hudson, and H.L. Falk. 1976. Embryonal neoplasms in the opossum: A new model for solid tumors of infancy and childhood. *Science* **193:** 328.

Kaneko, Y., M. Celida Egues, and J.D. Rowley. 1981. Interstitial deletion of short arm of chromosome 11 limited to Wilms' tumor cells in a patient without aniridia. *Cancer Res.* **41:** 4577.

Kaneko, Y., C. Homma, J. Hata, N. Maseki, and H. Miyoshi. 1991. Chromosome and molecular analysis of Wilms' tumor (WT). US-Japan Symposium on Cancers of the Kidney, Section G. Hawaii. (Abstr.)

Kaneko, Y., K. Kondo, J.D. Rowley, J.W. Moohr, and H.S. Maurer. 1983. Further chromosome studies on Wilms' tumor cells of patients without aniridia. *Cancer Genet. Cytogenet.* **10:** 191.

Kao, F.T., C. Jones, and T.T. Puck. 1976. Genetics of somatic mammalian cells: Genetic, immunologic, and biochemical analysis with Chinese hamster cell hybrids containing selected human chromosomes. *Proc. Natl. Acad. Sci.* **73:** 193.

Knudson, A.G., Jr. and L.C. Strong. 1972. Mutation and cancer: A model for Wilms' tumor of the kidney. *J. Natl. Cancer Inst.* **48:** 313.

Kondo, K., R.R. Chilcote, H.S. Maurer, and J.D. Rowley. 1984. Chromosome abnormalities in tumor cells from patients with sporadic Wilms' tumor. *Cancer Res.* **44:** 5376.

Koufos, A., M.F. Hansen, B.C. Lampkin, M.L. Workman, N.G. Copeland, N.A. Jenkins, and W.K. Cavenee. 1984. Loss of alleles at loci on human chromosome 11 during genesis of Wilms' tumour. *Nature* **309:** 178.

Koufos, A., P. Grundy, K. Morgan, K.A. Aleck, T. Hadro, B.C. Lampkin, A. Kalbakji, and W.K. Cavenee. 1989. Familial Wiedemann-Beckwith syndrome and a second Wilms' tumor locus both map to 11p15.5. *Am. J. Hum. Genet.* **44:** 711.

Kovacs, G., S. Szücs, and H. Maschek. 1987. Two chromosomally different cell populations in a partly cellular congenital mesoblastic nephroma. *Arch. Pathol. Lab. Med.* **111:** 383.

Kumar, S., C.J. Harrison, J. Heighway, H.B. Marsden, D.C. West, and P. Morris Jones. 1987. A cell line from Wilms' tumor with deletion in short arm of chromosome 11. *Int. J. Cancer* **40:** 499.

Ladd, W.E. 1938.. Embryoma of the kidney (Wilms' tumor). *Ann Surg.* **108:** 885.

Lavedan, C., F. Barichard, M. Azoulay, P. Couillin, D. Molina Gomez, H. Nicolas, B. Quack, M.-O. Rethoré, B. Noel, and C. Junien. 1989. Molecular definition of de novo and genetically transmitted WAGR-associated rearrange-

ments of 11p13. *Cytogenet. Cell Genet.* **50:** 70.
Lewis, E.B. 1950. The phenomenon of the position effect. *Adv. Genet.* **3:** 73.
Lewis, W.H., H. Yeger, L. Bonetta, H.S.L. Chan, J. Kang, C. Junien, J. Cowell, C. Jones, and L.A. Dafoe. 1988. Homozygous deletion of a DNA marker from chromosome 11p13 in a sporadic Wilms' tumor. *Genomics* **3:** 25.
Liban, E. and I.L. Kozenitzky. 1970. Metanephric hamartomas and nephroblastomatosis in siblings. *Cancer* **25:** 885.
Lubinsky, M., J. Hermann, A.L. Kosseff, and J.M. Opitz. 1974. Autosomal-dominant sex-dependent transmission of the Wiedemann-Beckwith syndrome. *Lancet* **I:** 932.
Manivel, J.C., R.K. Sibley, and L.P. Dehner. 1987. Complete and incomplete Drash syndrome: A clinicopathologic study of five cases of a dysonto-genetic-neoplastic complex. *Hum. Pathol.* **18:** 80.
Mannens, M., P. Devilee, J. Bliek, I. Mandjes, J. de Kraker, C. Heyting, R.M. Slater, and A. Westerveld. 1990. Loss of heterozygosity in Wilms' tumors, studied for six putative tumor suppressor regions, is limited to chromosome 11. *Cancer Res.* **50:** 3279.
Mannens, M., R.M. Slater, C. Heyting, J. Bliek, J. de Kraker, N. Coad, P. de Pagter-Holthuizen, and P.L. Pearson. 1988. Molecular nature of genetic changes resulting in loss of heterozygosity of chromosome 11 in Wilms' tumours. *Hum. Genet.* **81:** 41.
Mannens, M., E.M. Bleeker-Wagemakers, K. Bliek, J. Hoovers, I. Mandjes, S. van Tol, R.R. Frants, C. Heyting, A. Westerveld, and R.M. Slater. 1989. Autosomal dominant aniridia linked to the chromosome 11p13 markers catalase and D11S151 in a large Dutch family. *Cytogenet. Cell Genet.* **52:** 32.
Matsunaga, E. 1981. Genetics of Wilms' tumor. *Hum. Genet.* **57:** 231.
McCoy, F.E., Jr., W.A. Franklin, A.J. Aronson, and B.H. Spargo. 1983. Glomerulonephritis associated with male pseudohermaphroditism and nephroblastoma. *Am. J. Surg. Pathol.* **7:** 387.
McDowell, H., P. Howard, J. Martin, C. Hart, and J. Crampton. 1989. Chromosome 1 studies in Wilms' tumor. *Cancer Genet. Cytogenet.* **43:** 203.
McNeil, W.H., Jr., and A.J. Chilko. 1938. Status of surgical and irradiation treatment of Wilms' tumor and report of two cases. *J. Urol.* **39:** 287.
Michalopoulos, E.E., P.I. Bevilacqua, N. Stokoe, V.E. Powers, H.F. Willard, and W.H. Lewis. 1985. Molecular analysis of gene deletion in aniridia-Wilms' tumor association. *Hum. Genet.* **70:** 157.
Miller, R.W., J.F. Fraumeni, and M.D. Manning. 1964. Association of Wilms' tumor with aniridia, hemihypertrophy and other congenital malformations. *N. Eng. J. Med.* **270:** 922.
Mitchell, P.J. and R. Tjian. 1989. Transcriptional regulation in mammalian cells by sequence-specific DNA binding proteins. *Science* **245:** 371.
Moore, J.W., S. Hyman, S.E. Antonarakis, E.H. Mules, and G.H. Thomas. 1986. Familial isolated aniridia associated with a translocation involving chromosomes 11 and 22 [t(11;22)(p13;q12.20]. *Hum. Genet.* **72:** 297.
Nardelli, J., T.J. Gibson, C. Vesque, and P. Charnay. 1991. Base sequence discrimination by zinc-finger DNA-binding domains. *Nature* **349:** 175.
Nakagome, Y., T. Ise, M. Sakurai, T. Nakajo, E. Okamoto, T. Takano, Y. Nakahori, Y. Tsuchida, N. Nagahara, Y. Takada, Y. Ohsawa, S. Sawaguchi, A. Toyosaka, N. Kobayashi, E. Matsunaga, and S. Saito. 1984. High-resolution studies in patients with aniridia-Wilms' tumor association,

Wilms' tumor or related congenital abnormalities. *Hum. Genet.* 67: 245.
Nehlin, J.O. and H. Ronne. 1990. Yeast MIG1 repressor is related to the mammalian early growth response and Wilms' tumor finger protein. *EMBO J.* 9: 2891.
Niikawa, N., Y. Fukushima, N. Taniguchi, S. Iizuka, and T. Kajii. 1982. Chromosome abnormalities involving 11p13 and low erythrocyte catalase activity. *Hum. Genet.* 60: 373.
Niikawa, N., S. Ishikiriyama, S. Takahashi, A. Inagawa, H. Tonoki, Y. Ohta, N. Hase, T. Kamei, and T. Kajii. 1986. The Wiedemann-Beckwith syndrome: Pedigree studies on five families with evidence for autosomal dominant inheritance with variable expressivity. *Am. J. Med. Genet.* 24: 41.
Nisen, P.D., K.A. Zimmerman, S.V. Cotter, F. Gilbert, and F.W. Alt. 1986. Enhanced expression of the N-*myc* gene in Wilms' tumors. *Cancer Res.* 46: 6217.
Olcott, C.T. 1950. A transplantable nephroblastome (Wilms' tumor) and other spontaneous tumors in a colony of rats. *Cancer Res.* 10: 625.
Orkin, S.H., D.S. Goldman, and S.E. Sallan. 1984. Development of homozygosity for chromosome 11p markers in Wilms' tumour. *Nature* 309: 172.
Osada, I., Y. Shiroko, Y. Miyamoto, M. Sekiguchi, G. Fugi, M. Ogata, H. Watanabe, and N. Kamada. 1981. Establishment of a cell line (W-2) derived from human Wilms' tumor showing chromosome 11 short arm interstitial deletions. In *Proceedings of the Japanese Cancer Association. Abstract of the 40th Annual Meeting*, p. 81. Sapporo, Japan.
Oshimura, M., H. Kugoh, M. Koi, M. Shimizu, H. Yamada, H. Satoh, and J.C. Barrett. 1990. Transfer of a normal human chromosome 11 suppresses tumorigenicity of some but not all tumor cell lines. *J. Cell. Biochem.* 42: 135.
Pelletier, J., M. Schalling, A.J. Buckler, A. Rogers, D.A. Haber, and D. Housman. 1991a. The Wilms' tumor gene (*WT1*) is involved in genitourinary development. *Genes Dev.* 5: 1345.
Pelletier, J., W. Bruening, S.P. Li, D.A. Haber, T. Glaser, and D. Housman. 1991b. Genetic evidence implicating the Wilms' tumor gene (WT1) in genitourinary development. *Nature* 353: 431.
Pelletier, J., W. Bruening, C.E. Kashtan, S.M. Mauer, J.C. Maniuel, J.E. Striegel, D.C. Houghton, C. Junien, R. Habib, L. Fouser, R.N. Fine, B.L. Silverman, D.A. Haber, and D. Housman. 1991c. Germline mutations in the Wilms' tumor suppressor gene are associated with abnormal urogenital development in Denys-Drash syndrome. *Cell* (in press).
Pendergrass, T.W. 1976. Congenital anomalies in children with Wilms' tumor, a new survey. *Cancer* 37: 403.
Perlman, M., G.M. Goldberg, J. Bar-Ziv, and G. Danovitch. 1973. Renal hamartomas and nephroblastomatosis with fetal gigantism: A familial syndrome. *J. Pediatr.* 83: 414.
Pettenati, M.J., J.L. Haines, R.R. Higgins, R.S. Wappner, C.G. Palmer, and D.D. Weaver. 1986. Wiedemann-Beckwith syndrome: Presentation of clinical and cytogenetic data on 22 new cases and review of the literature. *Hum. Genet.* 74: 143.
Porteous, D.J., W. Bickmore, S. Christie, P.A. Boyd, G. Cranston, J.M. Fletcher, J.R. Gosden, D. Rout, A. Seawright, K.O.J. Simola, V. van Heyningen, and N.D. Hastie. 1987. HRAS1-selected chromosome transfer generates

markers that colocalize aniridia- and genitourinary dysplasia-associated translocation breakpoints and the Wilms' tumor gene within band 11p13. *Proc. Natl. Acad. Sci.* **84:** 5355.

Potter, E.L. 1972. *Normal and abnormal development of the kidney*, p. 72. Year Book, Chicago.

Pritchard-Jones, K., S. Fleming, D. Davidson, W. Bickmore, D. Porteous, C. Gosden, J. Bard, A. Buckler, J. Pelletier, D. Housman, V. van Heyningen, and N. Hastie. 1990. The candidate Wilms' tumour gene is involved in genitourinary development. *Nature* **346:** 194.

Puck, T.T., P. Wuthier, C. Jones, and F. Kao. 1971. Genetics of somatic mammalian cells: Lethal antigens as genetic markers for study of human linkage groups. *Proc. Natl. Acad. Sci.* **68:** 3102.

Raubitschek, H. 1912. Uder eine Bosartige Nierengeschwulst bei einem Kindlichen Hermaphroditen. *Frankf. Z. Pathol.* **10:** 206.

Rauscher, F.J. III, J.F. Morris, O.E. Toumay, D.M. Cook, and T. Curran. 1990. Binding of the Wilms' tumor locus zinc finger protein to the EGR-1 consensus sequence. *Science* **250:** 1259.

Reeve, A.E., S.A. Shih, A.M. Raizis, and A.P. Fienberg. 1989. Loss of allelic heterozygosity at a second locus on chromosome 11 in sporadic Wilms' tumor cells. *Mol. Cell. Biol.* **9:** 1799.

Reeve, A.E., M.R. Eccles, R.J. Wilkins, G.I. Bell, and L.J. Millow. 1985. Expression of insulin-like growth factor-II transcripts in Wilms' tumor. *Nature* **317:** 258.

Reeve, A.E., P.J. Housiaux, R.J.M. Gardner, W.E. Chewings, R.M. Grindley, and L.J. Millow. 1984. Loss of a Harvey *ras* allele in sporadic Wilms' tumour. *Nature* **309:** 174.

Riccardi, V.M., E. Sujansky, A.C. Smith, and U. Francke. 1978. Chromosomal imbalance in the aniridia/Wilms' tumor association: 11p interstitial deletion. *Pediatrics* **61:** 604.

Riccardi, V.M., H.M. Hittner, L.C. Strong, D.J. Fernbach, R. Lebo, and R.E. Ferrell. 1982. Wilms' tumor with aniridia/iris dysplasia and apparently normal chromosomes. *J. Pediatr.* **100:** 574.

Rinderknecht, E. and R.E. Humbel. 1976. Polypeptides with nonsuppressible insulin-like and cell-growth promoting activities in human serum: Isolation, chemical characterization, and some biological properties of form I and II. *Proc. Natl. Acad. Sci.* **73:** 2365.

Romain, D.R., O.B. Gebbie, R.G. Parfitt, L.M. Columbano-Green, R.H. Smythe, C.J. Chapman, and A. Kerr. 1983. Two cases of ring chromosome 11. *J. Med. Genet.* **20:** 380.

Rose, E.A., T. Glaser, C. Jones, C.L. Smith, W.H. Lewis, K.M. Call, M. Minden, E. Champagne, L. Bonetta, H. Yeger, and D.E. Housman. 1990. Complete physical map of the WAGR region of 11p13 localizes a candidate Wilms' tumor gene. *Cell* **60:** 495.

Schoenle, E., J. Zapf, R.E. Humbel, and E.R. Froesch. 1982. Insulin-like growth factor I stimulates growth in hypophysectomized rats. *Nature* **296:** 252.

Schofield, P.N., S. Lindham, and W. Engstrom. 1989. Analysis of gene dosage on chromosome 11 in children suffering from Beckwith-Wiedemann syndrome. *Eur. J. Pediatr.* **148:** 320.

Schroeder, W.T., L.-Y. Chao, D.D. Dao, L.C. Strong, S. Pathak, V. Riccardi, W.H. Lewis, and G.F. Saunders. 1987. Nonrandom loss of maternal

chromosome 11 alleles in Wilms' tumors. *Am. J. Hum. Genet.* **41**: 413.
Schroeder, W.T., L.Y. Chao, H. Farrel, B. Kikucli, W. Lewis, S. Nichols, V. Pothak, V. Riccardi, G.F. Saunders, and L. Strong. 1985. Relationship between human catalase activity and gene copy number. *Am. J. Hum. Genet.* **37**: A173.
Scott, J., J. Cowell, M.E. Robertson, L.M. Priestly, R. Wadey, B. Hopkins, J. Pritchard, G.I. Bell, L.B. Rall, C.F. Graham, and T.J. Knott. 1985. Insulin like growth factor-II gene expression in Wilms' tumor and embryonic tissues. *Nature* **317**: 260.
Scrable, H., W. Cavenee, F. Ghavimi, M. Lovell, K. Morgan, and C. Sapienza. 1989. A model for embryonal rhabdomyosarcoma tumorigenesis that involves genome imprinting. *Proc. Natl. Acad. Sci.* **86**: 7480.
Scully, R.E. 1970. Gonadoblastoma: A review of 74 cases. *Cancer* **25**: 1340.
Shanklin, D.R. and C. Sotelo-Avila. 1969. In situ tumors in fetuses, newborns and young infants. *Biol. Neonate* **14**: 286.
Shaw, M.W., H.F. Falls, and J.V. Neel. 1960. Congenital aniridia. *Am. J. Hum. Genet.* **12**: 389.
Sherwood, J.B., R. Bard, M. Bhargava, E.R. Burns, and M.A. Iqbal. 1989. A human adult Wilms' tumor. Histologic, ultrastructural, and cytogenetic analysis. *Cancer Genet. Cytogenet.* **42**: 35.
Simola, K.O.J., S. Knuutila, I. Kaitila, A. Pirkola, and P. Pohja. 1983. Familial aniridia and translocation t(4;11)(q22;p13) without Wilms' tumor. *Hum. Genet.* **63**: 158.
Slater, R.M. and J. de Kraker. 1982. Chromosome number 11 and Wilms' tumor. *Cancer Genet. Cytogenet.* **5**: 237.
Slater, R.M., J. de Kraker, P.A. Voûte, and J.F.M. Delemarre. 1985. A cytogenetic study of Wilms' tumor. *Cancer Genet. Cytogenet.* **14**: 95.
Slye, M., H.F. Holmes, and H.G. Wells. 1921. Primary spontaneous tumors in the kidney and adrenal of mice. Studies on the incidence and inheritability of spontaneous tumors in mice. *J. Cancer Res.* **6**: 305.
Solis, V., J. Pritchard, and J.K. Cowell. 1988. Cytogenetic changes in Wilms' tumors. *Cancer Genet. Cytogenet.* **34**: 223.
Sotelo-Avila, C., F. Gonzalez-Crussi, and J.W. Fowler. 1980. Complete and incomplete forms of Beckwith-Wiedemann syndrome: Their oncogenic potential. *J. Pediatr.* **96**: 47.
Spear, G.S., T.P. Hyde, R.A. Gruppo, and R. Slusser. 1971. Pseudohermaphroditism, glomerulonephritis with nephrotic syndrome, and Wilms' tumor in infancy. *J. Pediatr.* **79**: 677
Stiller, C.A., E.L. Lennox, and L.M. Kinnier Wilson. 1987. Incidence of cardiac septal defects in children with Wilms' tumour and other malignant diseases. *Carcinogenesis* **8**: 129.
Stump, T.A. and R.A. Garrett. 1954. Bilateral Wilms' tumor in a male pseudohermaphrodite. *J. Urol.* **72**: 1146.
Tank, E.S. and T. Melvin. 1990. The association of Wilms' tumor with nephrologic disease. *J. Pediatr. Surg.* **25**: 724.
Theiler, K., D.S. Varnum, and L.C. Stevens. 1978. Development of Dickie's small eye, a mutation in the house mouse. *Anat. Embrol.* **155**: 81.
Thomas, I.T. and D.W. Smith. 1974. Oligohydramnias, cause of the nonrenal features of Potter's syndrome, including pulmonary hypoplasia. *J. Pediatr.* **84**: 811.

Thompson, S.W., R.A. Huseby, M.A. Fox, C.L. Davis, and R.D. Hunt. 1961. Spontaneous tumors in the Sprague-Dawley rat. *J. Natl. Cancer Inst.* **27:** 1037.

Tomashefsky, P., Y.L. Homsy, J.K. Lattimer, and M. Tannenbaum. 1976. A murine Wilms' tumor as a model for chemotherapy and radiotherapy. *J. Natl. Cancer Inst.* **56:** 137.

Ton, C.C.T., V. Huff, K.M. Call, S. Cohn, L.C. Strong, D.E. Housman, and G.F. Saunders. 1991. Smallest region of overlap in Wilms' tumor deletions uniquely implicates an 11p13 zinc finger gene as the disease locus. *Genomics* **10:** 293.

Turleau, C., J. de Grouchy, F. Chavin-Colin, H. Martelli, M. Voyer, and R. Charlas. 1984. Trisomy 11p15 and Beckwith-Wiedemann syndrome. A report of two cases. *Hum. Genet.* **67:** 219.

van der Meer-de Jong, R., M.E. Dickinson, R.P. Woychik, L. Stubbs, C. Hetherington, and B.L.M. Hogan. 1990. Location of the gene involving the small eye mutation mouse chromosome 2 suggests homology with human aniridia 2 (AN2). *Genomics* **7:** 270.

van Heyningen, V., P.A. Boyd, A. Seawright, J.M. Fletcher, J.A. Fantes, K.E. Buckton, G. Spowart, D.J. Porteous, R.E. Hill, M.S. Newton, and N.D. Hastie. 1985. Molecular analysis of chromosome 11 deletions in aniridia-Wilms' tumor syndrome. *Proc. Natl. Acad. Sci.* **82:** 8592.

van Heyningen, V., W.A. Bickmore, A. Seawright, J.M. Fletcher, J. Maule, G Fekete, M. Gessler, G.A.P. Bruns, C. Huerre-Jeanpierre, C. Jurrien, B.R.G. Williams, and N.D. Hastie. 1990. Role for the Wilms' tumor gene in genital development. *Proc. Natl. Acad. Sci.* **87:** 5383.

Wales, J.K.H., V. Walker, I.E. Moore, and P.T. Clayton. 1986. Bronze baby syndrome, biliary hypoplasia, incomplete Beckwith-Wiedemann syndrome and partial trisomy 11. *Eur. J. Pediatr.* **145:** 141.

Wang-Wuu, S., S. Soukup, K. Bove, B. Gotwals, and B. Lampkin. 1990. Chromosome analysis of 31 Wilms' tumors. *Cancer Res.* **50:** 2786.

Watts, S.L. and R.E. Smith. 1980. Pathology of chickens infected with avian nephroblastoma virus MAV-2(N). *Infection Immunol.* **27:** 501.

Waziri, M., S.R. Patil, J.W. Hanson, and J.A. Bartley. 1983. Abnormality of chromosome 11 in patients with features of Beckwith-Wiedemann syndrome. *J. Pediatr.* **102:** 873.

Weissman, B.E., P.J. Saxon, S.R. Pasquale, G.R. Jones, A.G. Geiser, and E.J. Stanbridge. 1987. Introduction of a normal human chromosome 11 into a Wilms' tumor cell line controls its tumorigenic expression. *Science* **236:** 175.

Wiedemann, H.R. 1964. Complexe malformatif familial avec hernie ombilicale et macroglossie-Un syndrome noveau? *J. Genet. Hum.* **13:** 223.

———. 1983. Tumor and hemihypertrophy associated with Wiedemann-Beckwith's syndrome. *Eur. J. Pediatr.* **141:** 129.

Williams, J.C., K.W. Brown, M.G. Mott, and N.J. Maitland. 1989. Maternal allele loss in Wilms' tumor. *Lancet* **I:** 283.

Wilms, M. 1899. *Die Mischgeschwulste der Nieren*, p. 1. Arthur Georgi, Leipzig.

Wilson, J.D. and P.C. Walsh. 1979. Disorders of sexual differentiation. In *Campbell's urology*, 4th edition (ed. J.H. Harrison et al.), p. 1512. W.B. Saunders. Philadelphia.

Wolman, S.R., P.M. Camuto, A.J. Eisenberg, T.M. Guarino, and M.A. Greco.

1987. Wilms' tumor: A search for the critical lesion. *Am. J. Hum. Genet.* **41**: A41.

Yunis, J.J. and N.K.C. Ramsay. 1980. Familial occurrence of the aniridia-Wilms' tumor syndrome with deletion 11p13-14.1. *J. Pediatr.* **96**: 1027.

Index

Alternative splicing, 47

Beckwith-Wiedemann syndrome, 140–143
Breast tumors, *RB1* and, 8

c-Ha-*ras*, 141
Choroideremia, 2, 11–13
Chromosomal localization, diseases associated with
  choroideremia, 2, 11–13
  chronic granulomatous disease, 1, 3–5
  cystic fibrosis, 2, 9–11
  Duchenne/Becker muscular dystrophy, 1, 5–7
  neurofibromatosis type 1–2, 13–16
  retinoblastoma, 2, 7–9
  Wilms' tumor, 2, 16–17
Chromosome deletion syndrome, 12
Chromosome-mediated gene transfer, 10, 16, 148
Chromosome walking and jumping, 10, 12, 15, 17, 60, 64–67
Chronic granulomatous disease, 1
  b_{245} cytochrome, 5
  Duchenne muscular dystrophy patients with, 4
  positional cloning, 3–4
c-Ki-*ras*, 140
c-*kit*, 105–106
  chromosomal mapping of, 107–108
  expression, 111
    during embryogenesis, 122–126
  kinase activity, 111, 117
  mutations in mice, rats, and humans, 113, 124–125
Complementary cloning of Duchenne muscular dystrophy gene, 4
Complex disease studies
  genetic heterogeneity, 82–83
  high-resolution linkage mapping, 88–89
  identity by state, 89
  interval mapping, 83
  likelihood, 80–82
  linkage analysis
    in complex traits and λ_R, 86–88
    multiple marker testing, 91–92
  linkage disequilibrium, 83
  logistic function, 82
  penetrance and phenocopies, 82
  polymorphisms, 88
  segregation analysis, 84–86
Cystic fibrosis (CF), 83
  CF transmembrane conductance regulator, 10–11

chromosome walking and jumping, 10
linkage disequilibrium, 10
long-range physical mapping studies, 9
Cytochromes in chronic granulomatous disease patients, 4

Denys-Drash syndrome, 143–145, 155
Duchenne/Becker muscular dystrophy, 1, 4
  cross-species hybridization, 6
  X-autosome translocations, 5
Dystrophin, 7

Ectropic viral integration site-2 gene (*Evi-2*), 14
Epistasis, 80, 83–84
Esterase D, 7–8
Ethylnitrosourea, 156

Fibrillin, 19
Field inversion gel electrophoresis, 12
Flavoproteins in chronic granulomatous disease patients, 4

GTPase-activating protein, 15

HRAS1, 149
Huntington's disease, 20

Identity by descent, epistasis confounding linkage disequilibrium, 83–84
Insulin-dependent diabetes mellitus (IDDM), 79
  congenic mouse strains, 95, 98
  genetic analysis
    epidemiology, 92–93
    NOD mouse genetics, 93–95

immunogenetics, 94–95
microsatellite linkage map, 95–96
segregation analysis, 85

Kit receptor
  ligand-activated Kit receptor, cytoplasmic substrates, 117–118
  Sl encodes membrane-bound ligand for, 116–117
Kit signal transduction pathway, 105, 115–119
  in vivo functions
    gametogenesis, 121–122
    hematopoiesis, 119–120
    melanogenesis, 120–121

Linkage analysis, 3–4, 11, 13–14, 18
  in complex traits, 86–88
  identification of cystic fibrosis gene and, 2
Linkage disequilibrium, 10, 18, 83
  epistasis confounding linkage disequilibrium, 83–84
Long-range physical mapping studies and cystic fibrosis, 9

MAPMAKER, 83, 90
Mast cell growth factor, 116
McLeod syndrome, 4
Mesenchymal tumors, 8
MIC1, 149
MIC2, 62
  chromosome walking in, 65
microphthalmia (*mi*), 116
  locus required for Kit and FMS function, 118–120
MIG1, 153
myc, 145–146

Neurofibromatosis type 1 (NF1), 1–2, 19
  linkage analysis, 13–14
  NF1 gene isolation, 14–15
  as tumor suppressor gene, 16
  variable expressivity, 13

Osteosarcoma, RB tumorigenicity and, 8

Paraoxonase, 9
Patch, 123–124
Perlman syndrome, 145
Phagocytic cells in chronic granulomatous disease patients, 3–4
Phenol-enhanced reassociation technique (pERT), 4, 6, 18
Piebaldism, 125
Polymerase chain reaction (PCR), 20, 47, 72–73
Positional cloning, 60
  disease gene identification on the basis of chromosomal localization, 1–21
Pulsed-field gel electrophoresis, 9, 14, 16, 18–19
  Wilms' tumor analysis and, 149–151

Radiation-reduced somatic cell hybrids, 16
Receptor tyrosine kinases (RTKs), 106–107, 125–126
  and Kit signal transduction pathway, 115–119
  signaling pathways and development, 126
recessive spotting, 123–124
Restriction-fragment-length polymorphisms (RFLPs), 3, 6, 11, 88, 155
Retinitis pigmentosa, 4, 19
Retinoblastoma and *RB1* gene isolation, 8
  loss of heterozygosity, 8–9
  as recessive oncogene, 7–8
  two-hit hypothesis, 8–9
Reverse genetics, 60
  disease gene identification on the basis of chromosomal localization, 1–21
Rhodopsin, 19
Rump-white, 123–124

Saturation mapping technique, 10
sevenless, 126
Sex-determining gene (*TDF*), cloning of mammalian *TDF*
  chromosome walking, 64–65
  construction of transgenic mice, 74
  correlations between *SRY* and *TDF*, 69–74
  deletion fragment maps of Y-specific region, 62–64
  de novo *SRY* mutation, 72
  limiting step in positional cloning, 67–68
  role of pseudoautosomal (PAR) and Y-specific regions, 61
  long-range restriction map, 64
  MIC2, 62
  PAR meiotic map, 62–63
  role of testis and Y chromosome, 60, 70
  Tdy and, 61
  sex-reversed XX males, 62–64
  sex-reversed XY females, 64
  sry (in mice), 68, 70
  sex-reversed transgene, 74
  SRY, 68, 74
  mutation analysis in XY females, 70–74
  ZFY and, 66
  ZFY-negative males, 67
Sex determining region-*Y* gene (*SRY*), 68
Sey locus in mice, and Wilms' tumors, 156–157
Single-strand conformation polymorphism, 20
  assay, 70
Small cell lung carcinoma, 8
Somatic mutation model and retinoblastoma, 2
Sox, mouse autosomal gene, 70
Steel (*Sl*) mutation in mice, molecular biology
  Kit signaling pathway as product of *Sl*, 105
  Sl encodes membrane-bound ligand for Kit receptor, 116–117
Subtractive hybridization, 4

Tapetochoroidal dystrophy, 11–13
t complex responder (*Tcr*) locus in mice
  candidate gene, 46–50
  maps of mouse chromosome 17, 40
  t haplotypes, 39–42
  transmission ratio distortion (TRD) in mice carrying *t* haplotype
    genetics, 44–45
    model, 45, 50–53
    physiology, 42–44
    spermatogenesis, 42
Testis determining Y gene (*Tdy*), 61
Transgenic mice, 74
  as model for *Tcr* locus, 38–40, 48–50
Tumor suppressor gene experiments in Wilms' tumors, 145–146
Turner's syndrome, 61
Tyrosine kinase, 105. *See also* Receptor tyrosine kinases

Variable number of tandem repeats (VNTR), 3
von Recklinghausen disease. *See* Neurofibromatosis type 1

Waardenberg's syndrome, 86
WAGR syndrome, 136–140
White-spotting (*W*) mutation in mice, molecular biology, 110–111
  dominant-negative *W* phenotypes, 105
  hematopoietic defect in, 110
  melanogenesis, 120–121
  W/c-kit, 106, 109
    regulatory mutations, 111–112
    structural mutations, 114
Wilms' tumor (WT), 16–17
  11p13 WT gene, 147
    cell hybrids, 148–149
    mapping, 149–150
  Beckwith-Wiedemann syndrome, 140–143
  Denys-Drash syndrome, 143–145, 155
  IGF-II in WT, 141–143
  loss-of-heterozygosity studies, 146–147
  mouse *Sey* locus, 156–157
  Perlman syndrome, 145
  tumor suppressor gene experiments, 145–146
  WAGR syndrome, 136–140, 147
  WT1, 146
    chromosomal location, 152
    isolation and functional studies, 151–156

Yeast artificial chromosome (YAC), 15, 19, 65

Zinc finger domains, 17